The Mechanics of Physics

The Mechanics of Physics

A Rational Approach to Learning Newtonian Physics

by
Louis W. Bixby

VANTAGE PRESS
New York / Washington / Atlanta
Los Angeles / Chicago

FIRST EDITION

All rights reserved, including the right of
reproduction in whole or in part in any form.

Copyright © 1982 by Louis W. Bixby

Published by Vantage Press, Inc.
516 West 34th Street, New York, New York 10001

Manufactured in the United States of America
Standard Book Number 533-05117-7

Library of Congress Catalog Card No.: 81-90240

To the St. Louis Country Day School, a wonderful school, where these mechanics programs were actually used

Contents

Acknowledgments	ix
Introduction	xi
1. The Basic Schedule of Learning	1
2. The Basics of True Scientific Learning	3
3. Mechanical Rationale in Newtonian Physics	5
4. A Rational Approach to Learning Newtonian Physics	7
5. Newton's Laws of Motion	15
6. All Equations	18
7. Initial Experiments	24
8. Review Questions on Newtonian Mechanics	41
9. Newtonian Problems (Homework Items)	46
10. Vectors—Items to Analyze	51
11. Items on Momentum	57
12. Energy Items and Ideas	59
13. Newtonian Review	61
14. Quizzes and Tests	63
15. Newtonian Laboratory Items—Eleven Experiments	87

Acknowledgments

This basic science information has been developed stepwise in my many great schools with the instructive help of my great students. Sincerely, these students have had truly individualized learning steps, and therefore *true forward development*.

Eight Exciting Classrooms:

Dryden Central School, Dryden, New York
Willard Straight Hall, Cornell University
Groton Central School, Groton, New York
Reed College, Portland, Oregon
Weedsport Central School, Weedsport, New York
Saint Louis Country Day School, St. Louis, Missouri
Deerfield Academy, Deerfield, Massachusetts

and twenty-eight years of over 2,500 bright young humans!—Nary an acceptable dumb one in the lot—all of whom have kept this teacher *young*!

Introduction

This book is a *very basic* beginning step for students on the subject of *mechanics*. It is a set of items for the *easiest learning* in all steps with physics. The items are distance, time, and mass—the truly beginning ideas that all students know already, so they associate this as background for every other mathematically developed step. Therefore, learners who use these steps in the proper order will see a very *easy* aspect of true, *correct* learning!

Having understood the basic items for mechanics as indicated here, learners will certainly see that all other steps (light, sound, electricity, heat, and nuclear physics) have (practically) the *same* equations. This basic learning really adds up to much better, and *easier* learning in the rest of physics.

The beginning pages show the basics of true scientific learning, and then explain all points. The following pages illustrate a concise concentration of the basic points—the entire set of significant basic equations.

The use of this book will equal a very good set of individual steps for students, and it helps very much to bring about *true*, real learning steps. Related to such a "small" book, students can look through other, big physics books (in the library) and will see the association very well.

Related to this "small" book on mechanics, only related to physics, I do have equal student book knowledge of heat, light, and electricity. The "Mechanics, Only" approach adds up to a fine, exciting step for learners, and will then be a solid background for all other steps.

Items

Beginning talk:

Here, students, are a number of "ideas" of scientific knowledge. It relates to most of physics' ideas, which are really a most useful set of steps for you to know, and much of this adds up to many "thought through" science items on the observations. This honestly should show you how basic physics is *really needed* in education!

Here is the first science question for learners to analyze, and later see the many mechanical associations on such a simple item:

A number of books are sitting at rest on your desks. Which of the following statements best describe the situation?

A. There are forces acting on the paper.
B. There are forces acting, but they are balanced.
C. The paper-text exerts no force on the desk.
D. Your paper is at rest in any coordinate system.

Other Items:

1. A container of water sitting on the teacher's front area, for a month or two
2. Titanium Dioxide
3. Analysis of an egg while it is dropping
4. Other egg experiments—student egg-tossing
   This is to analyze as an idea, not as an immediate experiment!
5. Note the set of beginning physics here:
   Analysis of mechanics principles—(I have put this into my school-area, for students to always "see")
6. Basic colors—color analysis of light

Items 7, 8, and 9
(are for groups to work on, before we analyze anything)

7. Looking through the red window!
8. Jim Dandy X-Ray machine of light!

(AND—Maybe #2 here now)

9. Man, Seat, and Man *in* the Chair!
10. Mass being knocked forward
11. Different masses knocked forward
12. Force making an airplane rise

Items to Talk About

13. The bicycle laboratory—Again, we won't do this, but it is a great, fun lab for student riders to analyze.
14. A concave mirror—Again, this probably will not be here, but we can talk about it.

The Mechanics of Physics

1

The Basic Schedule of Learning

This basic handout of materials represents stepwise ideas and developments in our physics education. A number of other text references can be added (in the library), which you can use at your need and desire. The use of this book is mainly for the purpose of reference on what we analyze in lab and in class. You have received *many* learning ideas in the handout, which will in essence constitute your best text. Also note that there are included a number of TEST items for your practice and use, many of which will be test questions.

Activities for the Course

1. Keep a well-organized notebook with all handouts, class notes, and laboratory analyses.
2. Reading the student text and possibly other items
3. Problem solving (homework problems)
4. Development of equations and mathematical relationships.
5. Class discussions (on readings, lab work, and teacher *and* student ideas)
6. Quizzes—in class
7. Occasional Lectures by the teacher, and possibly some by students
8. References (slides, and such)
9. Oral examinations on subject areas studied - involving three or four students together with scientific ideas, on discussion and explanation.
10. Tests—in class and some take-home ones
11. Movies

12. Student Groups of three or four to demonstrate lab investigations
13. Laboratory Work—in class and under other circumstances. Much of the physics course will involve the use of the laboratory to develop and analyze physical principles.
14. Teacher demonstrations—for thoughts and analysis
15. Individual investigations

2

The Basics of True Scientific Learning

Science at its best and most exciting level is an approach whereby one can observe and learn in depth *on one's own*. The most famous and successful of scientists have been those who, in essence, sat back, used their brain power, and came forth with new and significant analyses on the observations of natural phenomena. We will discover and discuss examples of such people as Aristotle, Newton, Watson, and others. This list will include present scientists and people of centuries ago. In essence, the time for scientific discovery will never cease, for science is an ongoing theoretical view of the universe—not a factual absolute.

There are some *basics* to developing this individualized approach. One may think of physics as three coordinates—E—S—I, which mean nothing as individual letters, but everything as a workable scientific unit:

Energy   Structure   Interactions

One cannot describe energy in a void without relating to a structural entity. "Energy" is easy to *say* without really knowing what it means. Physical structures mean nothing significant unless one sees how they developed in their particular direction—an aspect of energy. These two basics, then, are the result of mutual *interactions*. A geometric view of a triangle shows that it can only be such a mathematical entity if all three points exist:

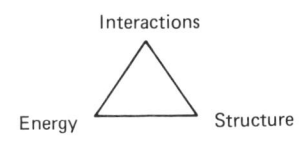

```
              Interactions
                  /\
                 /  \
                /    \
          Energy ---- Structure
```

Obviously, the diagram reflects what the teacher wants the student to do as far as of personal learning is concerned. Truly, students will develop more real understanding and thus, a most significant and exciting interest in science. If a person can truly teach himself, the teacher may well try hard at every point to not "teach" *per se*. The real role of a teacher can then be to "hold up the wall," if he sees that the so-called "students" only become true *learners* via individualized observation-digging on a subject. In physics, the "wall holding" capacity is much more than one might expect. Many experimental observations that are developed are concepts that one must personally analyze. Sometimes it is in relation to posed questions, and at other times, it is simply in relation to what one *thinks* it should convey based on knowledge and past experience. One must develop personal *gumption*.

An analogy to the original symbols **E, S, I** strongly says that there can and should be Interactions amongst Students and between Students and Educators, so—E S I! There will also be a number of demonstrations to be analyzed; test questions that will help the student relate to observations and do well with the other tests; printed outlines of significant scientific terms; relations; outlines of physics chapters; experimental suggestions; and much, much more. In using science books—it is important to learn to develop your own answers to interactions before reading the real, true explanation. With this in mind, the learner will develop and understand the course personally—*without* a text!

Related to the foregoing philosophy of true knowledge assimilation, here is the good set of potential learning steps:

Always Remember: The percentages of what *we receive* are 5 percent of what *we hear*, and 10 percent what *we see*. The percentages of what *we give back* are 20 percent of what *we echo* and 40 percent of what *we recite*.

The percentages of *our original activity*: are 50 percent of what *we read*, 60 percent of what *we discuss*, 75 percent of what *we create* and 95 percent of what *we **teach***.

So, are our students on the *receiving only* level, the *receiving and giving back* level, or the receiving and giving back original *activity* level? The answer to these potential steps gives us the answer to our *true* degree of real success!

3

Mechanical Rationale in Newtonian Physics

An aspect of teaching traditional physics is that a great number of equations are presented as *requirements* for memorization—seldom for true understanding. In many cases, these equations are learned without any *real* learning taking place! For instance, a great number of physics texts start (Chapter 1 or 2) with a variety of points, none of which are related to a truly rational, logical development of mathematical and physical steps. For instance, some books start with a discussion of force, and others on energy. Velocity often comes before speed, and so on. This adds up to hard learning difficulties, which can be simply resolved through considerations of how one truly looks at Newtonian nature. The question is, what do we really "see" physically? Our students will advocate that there really are such rational, structural steps.

Items that the beginning student can truly comprehend are *distance*, *time*, and *mass*. Our first step should be to start with these *only* observable realities of nature and build all other concepts around and upon them. Even distance, time, and mass are developments of our human look at nature, but they have the advantage of being the *simplest* one-unit points upon which all other mathematical views are built. Everything else, (speed, velocity, acceleration, force, momentum, inertia, vectors, weight, and energy) is a *derived* term. This small article to students is written to show how these derivations can easily and logically be developed in their minds.

The first mathematical step should be to put two of the fundamentals (distance, time, mass) into a meaningful unit. When observing anything that moves, we observe that it goes a certain distance in a given time. Our distance units are meters and time units are seconds and so forth.

"How far in what time?" translates to

$$\frac{\text{Distance}}{\text{Time}}$$

mathematically, which the Newtonian physicists called *speed*. (Distance per time makes sense—like miles per hour, whereas "distance x time" is a useless quantity.) Incidentally, speed in a specific *direction* (N, S, E, W,) is "vectorial," and is called *velocity*.

Then, it can be deduced that when velocity is changing, it changes as a unit of time, —and students/learners can logically develop an understanding of *acceleration*, mathematical step two.

Mass units are *kilograms* and *grams*, which are basically just quantities of matter. As the learner will see, the greater the mass of an object, the harder it is to move. Next, the *force* needed to move an object, relates velocity to mass. Just "playing" with heavy and light carts will show that the heavier (greater mass) cart will move slower when pushed—or will be harder to push if it moves at an equal velocity with the lighter cart. The resulting formula is

$$F = mv$$

Force equals mass times velocity, or kg times m/sec. This unit is called the *newtons* (one kg-m/sec) of force. The only items to be memorized at this point, should only be newtons, speed, and velocity. Only the newton is a completely new item to the learner. Everything else—the development of steps—should *make sense* after following a logical learning pattern.

4

A Rational Approach to Learning Newtonian Physics

Basic Units

Items that can be truly understood are distance, time, and mass (mass is probably the least understood). Our first step is to start with *only* these observable realities of nature, and build all other concepts around and upon them. Even distance, time, and mass are developed as humans look at nature. These are *simplest* one-unit points upon which all other mathematical views are built. Everything else—speed, velocity, acceleration, force, momentum, inertia, vectors, weight, and energy—is a *derived* mathematical term. This learning step is written to show how these derivations can most easily and logically be developed for the student's good understanding and use.

Speed and Velocity

The first mathematical step is to put two of the fundamentals (distance, time, mass) into a meaningful unit. When we see anything move, we observe that it goes a certain distance in a given time. Distance units are *meters*, and time units are *seconds* and so forth. "How far in what time?" mathematically translates to

$$\frac{\text{distance}}{\text{time}}$$

which the Newtonian physicists called *speed*. Speed in a specific *direction* (N, S, E, W, up, down, and so on) is *vectorial*, and is called "velocity."

Changing Velocity and Acceleration

Another calculation that is related to velocity is *average velocity*, designated as V. If velocity changes constantly, the V average, **V**, is represented as

$$\frac{V_{final} - V_{initial}}{2}$$

If $V_{initial}$ is zero, then

$$V = \frac{V_{final}}{2}$$

When velocity is changing, it changes in time (a fundamental) and one can then logically develop an understanding of *acceleration*, which is mathematical step two in our development. Velocity, the first mathematically derived concept, develops into *changing* velocity, where the change is in time. So the formula for changing velocity is

$$\text{velocity}, \frac{d}{t}, \text{times } \frac{1}{t}, \text{ and equals } \frac{d}{t^2}.$$

This is *acceleration*, usually noted in meters per second-squared, written as "m/sec². " Therefore

$$\frac{d}{t} \times \frac{1}{t} = \frac{d}{t^2} \text{ or } \frac{m}{\text{sec}} \times \frac{1}{\text{sec}} = \frac{m}{\text{sec}^2} = \frac{\text{distance}}{\text{time}^2}$$

Note that only one little thing is added, in a sense, to velocity, $\frac{1}{t}$.

Force

Where does *mass* come in? Units used are the kilogram and gram which are basically just a "quantity of matter." You should see that the greater the mass of an object, the harder it is to move. Thus, *force* is needed to move an object, which relates velocity to mass. Just "playing" with heavy and light carts, will show that the heavier (greater mass) cart will move slower when pushed—or will be harder to push if it moves at an equal velocity with the lighter cart. The resulting formula is

$$F = ma$$

Force equals mass multiplied by acceleration, or kg m/sec^2. This is called a *newton* (one kg-m/sec^2) of force. The items to be memorized at this point should only be newtons, speed, and velocity. Only the newton is a completely new item to the learner. Everything else—the developmental steps—should *make sense* following an easy, logical learning pattern. Furthermore, everything happens in units of time—so all calculations are *per* time; that is

$$\text{distance per time} = d/t$$
$$\text{velocity per time} = v/t$$
$$\text{acceleration per time} = a/t$$

Force exerted within a gravitational pull of the earth is called *weight*. Students can experiment and directly determine the acceleration due to gravity (gravitational force). It is 9.8 m/sec^2, called g. With this information, the gravitational force or weight of a one kilogram of mass is 9.8 newtons (9.8 kg-m/sec^2).

$$F = ma = 1 \text{ kg} \times 9.8 \text{ m/sec}^2 = mg = G \text{ (force of gravity)}.$$

Momentum and Inertia

Another consideration that relates mass to movement is *momentum*. The greater the mass an object has, or the faster it is moving, the larger is its momentum. Momentum thus equals mass times velocity, **mv**. A change (Δ) in momentum is mass times the change in velocity, mΔV; read "m-delta-V." Another aspect of momentum is *inertia*, or inertial mass, which indicates that for any two items moving at the same velocity, the one with the greatest mass will have the greatest *inertial mass*.

Summary of Motion

Steps considered thus far are:

distance and time
speed (the relationship between distance and time)
acceleration (changes in speed)

vector ideas (turning speed into velocity)
velocity
mass
force
weight or gravitational force
momentum and inertia

Work

The foregoing background serves as a foundation for understanding *work* (W), which happens when a force acts over a distance. So

$$W \quad Fd \quad (\text{Work} = \text{Force times distance}).$$

Work is a type of energy called *mechanical* (Newtonian mechanics). It serves well, developed logically as **Fd**, to introduce the energy concept. Students can see that no work is done on an object (no matter how massive) when it is stationary. Neither is work done if the object is moving in frictionless, non-gravitational space (a theoretical situation) at constant velocity. The next logical step in our thinking is to see that the object will *not* be able to do work on something else (move it due to a force Fd) unless it is accelerated or *de*celerated, (which is simply negative acceleration). Mathematical proof of this can be shown by playing with W = Fd in the following manner

$$W = Fd, \text{ the learning "definition" of work}$$
$$\text{Work} - \text{Force} \times \text{distance}$$
and $\quad F = ma \text{ (Force = mass} \times \text{acceleration)}$
So $\quad W \text{ must} = ma \times d \quad \text{or} \quad W = mad$

In other words, for the object to do work on something else, its mass must be *accelerated* (not just at constant velocity) over a distance of m a d.

Energy

A standard formula for energy is $E = \frac{1}{2}mv^2$. Again, one can logically relate this to the mathematical concept of mechanical energy—W = Fd—

according to the following:

Remember that $F = ma$,
So $W = mad$
and $\text{acceleration} = \dfrac{\text{distance}}{\text{time} \times \text{time}}$ or $a = \dfrac{d}{t^2}$

Also, $\text{velocity} = \dfrac{\text{distance}}{\text{time}} = \dfrac{d}{t}$

So

$\underset{(1)}{m} \quad \underset{(2)}{a} \quad \underset{(3)}{d} \quad = \quad \underset{(1)}{\text{mass}} \quad \times \quad \underset{(2)}{d/t^2} \quad \times \quad \underset{(3)}{\text{distance}}$

And $v^2 = \left(\dfrac{d}{t}\right)^2$ or $\dfrac{d^2}{t^2}$

which is $\dfrac{d}{t^2} \times d$

So, finally

a x d of mad equals v^2,

which gives us mv^2, above. Again

$E = \tfrac{1}{2} m v^2$ (or $\dfrac{mv^2}{2}$).

The ½ tells us that the acceleration rate will give an *average* velocity for any force ma, that acts through a distance on another object. (It could also be written as: $E = mv^2$.)

Potential and Kinetic Energy

Finally, the concept of *available* versus *used* energy for a system, comes out as *potential energy*, E_p, and *kinetic energy*, E_k. This is seen most easily in a graphical sense, that relates a 10 kilogram mass 100 meters "up" (in relation to the Earth's surface) to what its energy will be as it comes down. This represents a good method for review of Newtonian ideas, by which students can relate time, distance, velocity, acceleration, force, and so on to potential and kinetic energy values.

On the next page is a graph that involves the foregoing ideas. There is also a table of correlated calculations. You should now be able to analyze the graph and understand all the calculations. Later, with this kind of basic understanding, one can talk about potential energy in relation to heat reactions, energy of chemical systems, light, and so forth.

Also note the four pages of outlined equations. They are in "learning order" and for your understanding, not memorization *per se*.

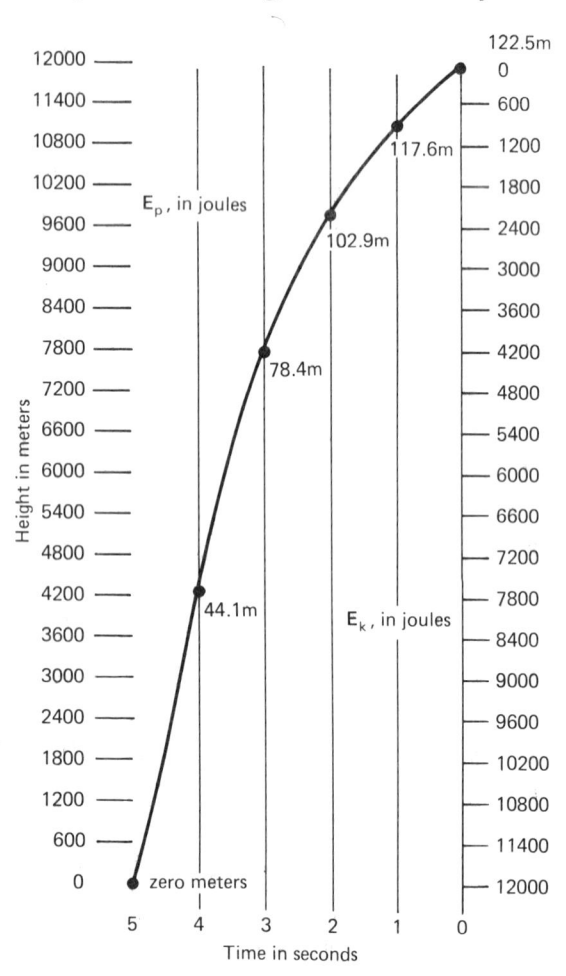

Correlated Calculations that Yield the Graphical Analysis:

Distance traveled, in meters per second	0	4.9	14.7	24.5	34.3	44.1
Time, in seconds	0	1	2	3	4	5
Final Velocity: m/sec	0	9.8	19.6	29.4	39.2	49.0
Initial Velocity: m/sec	0	0	9.8	19.6	29.4	39.2
Acceleration: m/sec^2	0	9.8	9.8	9.8	9.8	9.8
Force (ma), in newtons	0	9.8	9.8	9.8	9.8	9.8
Average Velocity: m/sec	0	4.9	14.7	24.5	34.3	44.1
Total Distance traveled, in meters	0	4.9	19.6	44.1	78.4	122.5
Total E_k, in joules	0	480.2	1920.8	4321.8	7683.2	12005.0
Total E_p, in joules	12005	11524.8	10084.2	7683.2	4321.8	0
Work (Fd) in one second	0	480.2	1440.6	2401.0	3361.4	4321.8

Physical Science with a Newtonian Background

The foregoing Newtonian view is an initial approach for understanding all other aspects of physics and other sciences. The basic sustenance for conceptualizing all other aspects of science is to firmly know what is meant by *energy*. The forms of energy—heat, light, sound, magnetic, electrostatic, and others, will logically develop through this understanding. And from energy considerations comes the true knowledge of structure and mass interactions such as elements, molecules, crystals, polymers, and living systems. Our most basic aim is to give the student a sound foundation to truly know energy by this very basic, logical, and relatively easily conceived development.

Related to these basic ideas, there are three laws that Newton developed. The copies of many different statements should well indicate that the individual learner can understand them through his or her own correct word analysis.

5

Newton's Laws of Motion

These are statements derived from nine textbooks on Newton's three laws.

First Law

A body remains in a state of rest or constant velocity (zero acceleration) *when left to itself; the net force acting on it is zero.*

If there is no net force on a body, it will continue in its state of rest or will continue moving along a straight line, with uniform velocity. Every particle continues in a state of rest or in uniform motion in a straight line unless it is compelled to change that state by an impressed force.

The Law of Inertia states that every body continues in its state of rest or uniform motion in a straight line unless it is compelled to change that state by forces impressed upon it. Any object that is at rest will remain at rest unless acted upon by a force. Any object in motion will continue in motion, with the same velocity and in the same direction, unless acted upon by a force.

The velocity of a body remains constant unless acted on by an outside, unbalanced force. If the net force acting on an object is zero, then the acceleration of the object is zero and it moves with constant velocity.

In symbols, $F = 0$ implies $A = 0$ or $v =$ constant

A body continues in its state of rest, or uniform motion in a straight line, unless it is acted upon by a net external force. A body remains at rest, or if in motion, it remains in uniform motion with constant speed in a straight line, unless it is acted on by an unbalanced external force.

Second Law

The net force on a body is the product of the mass of the body times its acceleration.

The effect of an applied force is to cause the body to accelerate in the direction of the force. The acceleration is in direct proportion to the force and in inverse proportion of the mass of the body.

A particle of mass m, subjected to a net force F, will experience an acceleration a that is proportional to the force and inversely proportional to the mass:

$$a = F/m.$$

Effect of a Force

Change of motion is proportional to the impressed force and takes place in the line in which the force acts. The observed acceleration of any object is directly proportional to the net applied force, and inversely proportional to the mass of the object. When an unbalanced force acts on a mass, it produces an acceleration in the direction of the force, that is directly proportional to the force and inversely proportional to the mass. A *force* is any influence that can cause a change in the state of motion of an object.

When an unbalanced force acts on a body, the body will be accelerated. The acceleration produced by an unbalanced force acting on a body is proportional to the magnitude of the net force, in the same direction as the force, and inversely proportional to the mass of the body.

Third Law

Whenever two bodies interact, the force on the first body due to the second is equal and opposite to the force on the second due to the first.

Whenever one body exerts a force on another, the second body exerts on the first a force of equal magnitude in the opposite direction. All forces arise from the mutual interactions between particles, and the force exerted by one particle on another is equal in magnitude but oppositely in direction to the force exerted by the second particle on the first. If one object applies a force to another, the second object applies an equal and opposite force to the first object. Thus, *For every action there is an equal and opposite reaction.*

Interactions

Whenever one object exerts a force on a second object, the second exerts an opposite force of equal magnitude on the first. Newton's words were, "To every action there is always opposed an equal reaction; or the mutual actions of two bodies upon each other are always equal and directed to contrary parts."

In nature, forces always come in pairs. These pairs are equal in magnitude but exactly opposite in direction. If one member of the pair is caused by object A and acts on object B, the other member of that pair will be the force due to object B acting on object A.

If we observe a force on an apple that is due to the gravitational attraction between the apple and the earth, we must also observe a force of exactly the same magnitude, but opposite in direction acting on the earth. Similarly, if object 1 exerts a force on object 2, then object 2 exerts an equal force, in an opposite direction, on object 1.

Law of Action and Reaction: Every force is accompanied by an equal and opposite force.

Whenever one body exerts a force upon a second body, the second body exerts a force upon the first body; these forces are equal in magnitude and opposite in direction.

6

All Equations

I. Velocity

d = distance   m = meters   t = time   sec = seconds

v = velocity = instantaneous velocity $v = \frac{d}{t} = \frac{m}{sec}$

Δ = change (delta)

Δd = change in distance = $d_f - d_i$

d_f = final distance, d_i = initial distance (or d_o)

Δt = change in time = $t_f - t_i$

t_f = final time, t_i = initial time (or t_o)

v_{av} = average velocity

$$v_{av} = \frac{d_f - d_i}{t_f - t_i} = \frac{\Delta d}{\Delta t}$$

II. Graphic Interpretation of Velocity

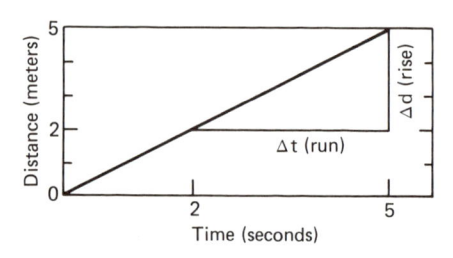

Δd (rise) = 5 m − 2 m
Δt (run) = 5 sec − 2 sec

$$Slope = \frac{\text{rise}}{\text{run}} = \frac{\Delta d}{\Delta t} = \frac{3 \text{ m}}{3 \text{ sec}} = 1 \text{ m/sec} \quad \text{(one meter per second)}$$

III. Change in Velocity—Acceleration

a = acceleration

Δv = change in velocity

$$a = \frac{\Delta v}{\Delta t} = \frac{\Delta \text{m/sec}}{\Delta \text{sec}} = \Delta \text{m/sec}^2$$

Constant acceleration = a = velocity increase (or decrease), Δv, by the same amount in each constant time (sec) interval.

$$a = \left(\frac{\text{change in instantaneous velocity}}{\text{change in time}} \right) = \left(\frac{\text{final instantaneous velocity} - \text{initial instantaneous velocity}}{\text{final time point} - \text{initial time point}} \right)$$

$$a = \frac{\Delta v}{\Delta t} = \frac{v_f - v_i}{t_f - t_i} = \frac{(\text{m/sec})_f - (\text{m/sec})_i}{(\text{sec})_f - (\text{sec})_i}$$

IV. Graphic Interpretation of Acceleration

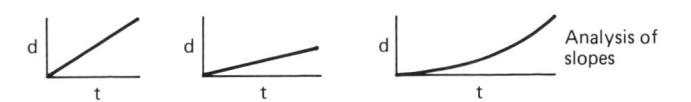

Analysis of slopes

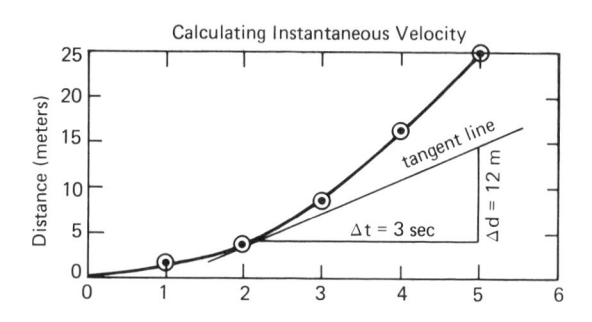

[Graph: Calculating Instantaneous Velocity — Distance (meters) vs time, showing tangent line with Δt = 3 sec and Δd = 12 m]

$$\frac{\Delta v}{\Delta t} = \frac{m/sec}{sec} = a$$

$\Delta v \times \Delta t = d$ = area under a velocity/time curve

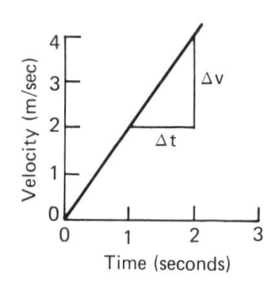

[Graph: Velocity (m/sec) vs Time (seconds), showing Δv and Δt]

V. Force

f = force

nt = newton = 1 kg-m/sec^2
m = mass = ratio of force to acceleration = *inertial* mass

kg = kilogram   g = gram

$$M = \frac{f}{a} = \frac{nt}{m/sec^2} = \frac{kg\text{-}m/sec^2}{m/sec^2} = kg$$

So: $F = ma = kg \times m/sec^2$

VI. Mass and Weight

W = weight = *force* with which an object is attracted to the Earth
g = acceleration due to gravity = 9.8 m/sec^2
W = Kg x 9.8 m/sec^2 = 9.8 nt
   (A mass of 1 Kg weighs 9.8 nt at sea level on earth.)

VII. Interrelationships between Calculations (dimensions) of Velocity, Acceleration, and Force

Force = F = newtons = kg-m/sec^2 or   kg-m/sec/sec

$$a = \frac{f}{m} = \frac{kg\text{-}m/sec^2}{kg} = \frac{kg\text{-}m}{sec^2} \times \frac{1}{kg} = \frac{m}{sec^2}$$

$$v = a \times t = \frac{m}{sec^2} \times sec = \frac{m}{sec}$$

or

$$\frac{Force}{time \times mass} = \frac{nt}{sec \times kg} = \frac{ma}{sec \times kg} = \frac{kg\text{-}m/sec^2}{sec \times kg} = \frac{m}{sec}$$

VIII. Impulse and Momentum

Impulse = force per unit time

$$= F \Delta t$$

$$a = \frac{\Delta v}{\Delta t}$$

$$F = ma = m \frac{\Delta v}{\Delta t}$$

So: FΔt = mΔv

Momentum = mΔv   (change in momentum)

   Initial or final momentum = mv

IX. Work and Energy

W = Work
W = Force x distance = Fd
  = newtons x meters
  1 joule = 1 nt-m
Force = mass x acceleration, F = ma
newton = nt = Kg-m/sec^2

So

Work = mass x acceleration x distance
W = mad
Units for a x d:

$$\frac{m}{sec^2} \times m = \frac{m^2}{sec^2} = ad$$

W = mad = Kg-m^2/sec^2 a x d also = v^2

E_k = Kinetic Energy (or KE), the energy of motion
The total E_k for a mass starting at rest, accelerating constantly, and ending at velocity, v, is related to the average velocity V_{av} $V_f/2$

So: $E_k = m \frac{v^2}{2} = \frac{1}{2}mv^2$ and   ½Kg-(m/sec)^2 = ½ joule

X. Potential Energy

E_p = potential energy (or **PE**)
Work = change in kinetic energy   W = ΔE_k = Fd = $\Delta \frac{1}{2}mv^2$
ΔE_p = mg (Δh)   h = height
Where the change in E_k = zero (no change in velocity), the work change in E_p is

W = ΔE_p = mgΔh = Kg x 9.8 m/sec^2 x Δm

XI. Conservation of Energy

W = ΔE_p and   W = ΔE_k

$E_p + E_k$ = constant $\Delta E_p + \Delta E_k = 0$

So:
$$-\Delta E_p = +\Delta E_k \quad \text{or} \quad +\Delta E_p = -\Delta E_k$$

And
$$W = \Delta E_k = \Delta E_p$$

Total work done by friction, kinetic energy and potential energy:

$$\begin{bmatrix} \text{Work done} \\ \text{by} \\ \text{Applied Force} \end{bmatrix} = \begin{bmatrix} \text{Work done} \\ \text{against} \\ \text{Friction} \end{bmatrix} + \begin{bmatrix} \text{Change in} \\ E_k \end{bmatrix} + \begin{bmatrix} \text{Change in} \\ E_p \end{bmatrix}$$

7

Initial Experiments

As an initial look at Newtonian physics, you will undergo a number of observations related to moving and interacting masses. For each item, you should take data on the mass of the object moved, the distance it was moved, and the time that it took to move it as well as any other data you can put together from Newtonian information. If two objects interact in the experiment, analyze both together. For each set-up, data will have to be taken and the results must be graphed.

Time is always on the **X** axis, and distance is on the **Y**. Always record the units used.

Some graph lines can be together. For instance, with Experiment 3, put it all on one graph with varying masses indicated. After graphing the results, write a short set of conclusions that are associated such as distance, time and mass relationships. Also, for each graph, show time/distance relationships at given distances—perhaps every 10 or 20 centimeters. (This will take some *practice*, so do it several times working in pairs with a timer and meter stick.)

Titles of Experiments:

1. Mechanics of Water Molecules
2. Work Done by a Falling Hammer
3. Collisions in Two Dimensions
4. A Collision in Two Dimensions

5. Gun Muzzle Velocity
6. The Spring—Shooting a Ball Forward
7. Analysis of Newtonian Physics with the Whirligig
8. The Bicycle Experiment
9. Falling Bodies and the Swinging Pendulum

Not all of these labs will be required, probably, for write-up or to be turned in. Some will be demonstrations and others will be for group discussions and analysis.

Note also, with these initial experiments, that a number of items do not show a great amount of the item's steps. With some of these, students can individually develop the experiment—based on their physical knowledge. (In this book, such small items mostly have 'bigger' information to the teacher.)

Experiment 1—Mechanics of Water Molecules

For this experiment, you will observe the demonstration on what takes place with one plastic container. (Materials: water with container, plastic containers with pinholes, timer.) From this observation, re-do this one and do two others. Then develop as much as possible on aspects of molecular motion related to Newtonian principles.

Experiment 2—Work Done by a Falling Hammer

Introduction

To drive a nail into a board or a sharpened post or into firm earth, requires large forces. These forces are obtained by using hammers, sledges, or pile drivers. The hammer head or other heavy body is given a high velocity in the direction that the nail or post is to be driven. The resistance of the wood or earth is a large force that quickly reduces the velocity of the hammer to zero. Meanwhile, the moving hammer exerts an equal force on the nail or post, which causes penetration.

Concerning the nail diven by a falling hammer, in this experiment, questions such as these will arise:
What is the magnitude of the force exerted on the nail?
With what velocity does the hammer strike the nail?
How long does the impulsive force act on the nail?
How long does it take the nail to stop the hammer?
How much work is done in driving the nail?

Object

To determine answers to the above questions, and others that occur, using the laboratory model of a pile driver.

A. To Find the Force During Impact:

Work Done = Average force on nail X Penetration during impact.

Energy of hammer = Work done in lifting = Weight of hammer X Height lifted

Average force on nail X Penetration = Weight of hammer X Height lifted

$$(1)\ Fd = mgh$$

Manipulation: Place the nail in a hole drilled in the given block of wood to insure that the nail will be vertical. Place the block so that the head of the nail is directly under the center of the hammer when it strikes. Plan to drop the hammer the same distance each time. The distance fallen is measured by means of the adjustable meter stick. Before dropping the hammer, measure the overall height of block and nail with vernier calipers. Since the nail moves a very short distance at each blow, the combined height of block and nail should be measured to a hundredth of a centimeter. This measurement, to be made after each blow, is recorded as the "caliper reading."

After the hammer has been allowed to drop and strike the nail, remeasure the combined height of the nail and block and calculate the "penetration." Then, *reset* the meter-stick measurement for a second blow. Continue in this manner until the nail is driven in up to its head or until at least ten blows have been struck.

B. To Find the Velocity of Striking:

Assume no loss of energy as the hammer falls. Its kinetic energy just before striking equals its potential energy at its highest point.

(2) $\tfrac{1}{2}mv^2 = mgh;\quad v^2 = 2gh;\quad v = \sqrt{2gh}$

C. To Find the Time of Impact:

If the units of length, mass, time, force—write as: $F \times t = m \times v$

(3) $t = mv/F$

Calculate, by this equation, the time required to stop the hammer at the first and at the last blow.

Mass (m) of hammer: Weight (W) of hammer:
Height (h) of fall (the same for all blows):
Potential Energy of Lifted Hammer (mgh):
Velocity (v) of Striking:
Time required to stop hammer—First Blow: Last Blow:
Sample Data Outline, on the *Falling Hammer*

No. of Blows	Caliper Reading	Successive Penetrations	Force on Nail
1			
2			
3			
etc.			
10			

Suggested Questions for Study and Oral Report:

1. Is "the force of a blow" determined entirely by the moving object that strikes the blow, or does it also depend on the force with which the obstacle is able to oppose the motion?

2. One sometimes reads such a statement as "The gun can throw a shell that will strike a blow of 8 tons." Why does this statement have no definite meaning?

3. As the nail goes deeper would you *expect* the friction force between nail and wood to become any more or less? Should you therefore *expect* the force exerted by the hammer, which drives the nail, to become any more or less?

4. How do the results of this study furnish justification for the use of rubber heels on shoes, of ball catchers' mitts, and of sawdust for high jumpers to land on?

Experiment 3—Collisions in Two Dimensions: Momentum

This laboratory exercise is completely outlined because the basic steps are difficult to perfect. It involves collisions between two wooden balls. Materials to be used are:

Two wooden balls, bricks for stopping balls, and distance measurement; one cart with spring, to be used as the accelerating force.

The factors to be determined are simpler than what you have done with the three balls before. This will not be an exact head-on collision, as noted by the diagram below. Utilize the initial idea suggested in the following diagram to institute a number of collisions between the balls and devise a way of calculating momentum changes for the system.

Diagram:

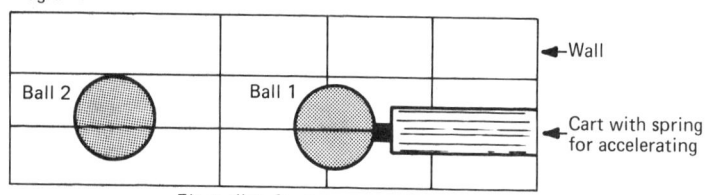

Floor tiles, for measurement

Ball 1 is to be accelerated by the cart's spring.
Ball 2 is "off center" from the path of motion of Ball 1 and is acted upon by Ball 1.
The floor tiles act as "graph paper" for analyzing the interaction.

The following is a report of experimental results and conclusions derived from data taken by a science teacher. It will, I hope, serve as a model for reporting on this experiment and doing the work.

The object of this experiment is to calculate the total momentum before and after the two-dimensional collision and to evaluate this in terms of energy conservation laws. The method employed to execute the experiment was to place a circle of bricks as a "backstop" around the approximate spot(s) at which the two balls would be expected to collide simultaneously. By moving this circle slightly through several successive trials, equal "momental distances" were determined. Points of initial and final collision were marked and the distances and angles between lines of displacement were measured.

The data from two trials indicates that the velocity of each ball *after* collision can be calculated. To do this, we first found two points that the ball hit instantaneously. We then measured back to the point of collision and took this distance measurement as a number from which to calculate relative velocities. (The time was equal for each ball to go from point of impact to point of stopping at the brick, so this time interval can be considered as unity—one second for convenience.) Our data tables indicate the magnitude of the velocities, but not their direction. The direction (and magnitude, to scale) is shown in Figure 1.

Initially, it seemed as though we could verify the law of conservation of momentum with one trial. However, we had no way of calculating the initial momentum of the first moving ball (Ball 1). Working vectorially "backward," from the first trial with the momenta of two balls, we could calculate initial momentum of the ball as it underwent an impressive force. This calculation *assumed* that the law of conservation of momentum *was* valid. The momentum of Ball 1, first trial, is shown as a negative sum of the two vectors in Figure 1, Figure 2.

On the second trial, we used the same ball (Ball 1) in initial motion but made the collision different by setting the struck ball (Ball 2) to the left of the line of collision instead of to the right, as in the first trial. With new collision positions, the direction of travel was slightly different and therefore the momenta was different. We attemped to measure the angle and distance traveled, both of which were close to the first trial angles and distances. If momentum is truly conserved, the sum of momenta in this trial should equal the sums of the momenta in the first trial. Therefore, the initial momenta (before collision) were equal and opposite to the final momenta.

Figure 3 shows the magnitude and direction of the two balls after collision. Their vector sum is also shown in the figure, and, within the limits of experimental error, checks with our calculations in Trial 1. In order to make a final check of our findings, we measured to scale a momentum equal to the calculated momentum of Ball 1 before collision and found its vector sum on the diagram, Figure 3. This was checked out experimentally and it was found that the distances traveled did obey the law of conservation according to our calculation and within limits of experimental error. Thus, in a two-dimensional collision, momentum appears to be conserved and the Law of Conservation of Momentum is validated.

Table 1

Trial	Ball 1 Mass (grams)	Ball 2 Mass (grams)	Ball 1 Distance traveled (cm.)	Ball 2 Distance traveled (cm.)	Ball 1 Velocity after collision (cm/sec.)	Ball 2 Velocity after collision (cm/sec.)
First	262.5	282.0	48.4	39.1	48.4	39.1
Second	262.5	282.0	42.6	46.8	42.6	46.8

Table 2
Calculations of Momenta before and after Collisions.
(The direction is omitted in this Table, but shown in Figure 4.)

	Ball 1	Ball 1	Ball 2	Ball 2
	momentum before collision–mv 262.5 x 59.6 = gms cm/sec 15,645.0 gm/cm/sec	momentum after collision–mv 265.5 x 48.4 = gms cm/sec 12,705 gm-cm/sec	momentum before collision–mv 282.0 x 0 = gms cm/sec 0 gm-cm/sec	momentum after collision–mv 282.0 x 39.1 = gms cm/sec 11,026.2 gm-cm/sec

Trial 1 –above

262.5 x 62.0= gms cm/sec 16,275.0 gm-cm/sec	262.5 x 42.6 = gms cm/sec 1,182.5 gm-cm/sec	282.0 x 0 = gms cm/sec 0 gm-cm/sec	282.0 x 46.8 = gms cm/sec 13,197.6 gm-cm/sec

Figure 1

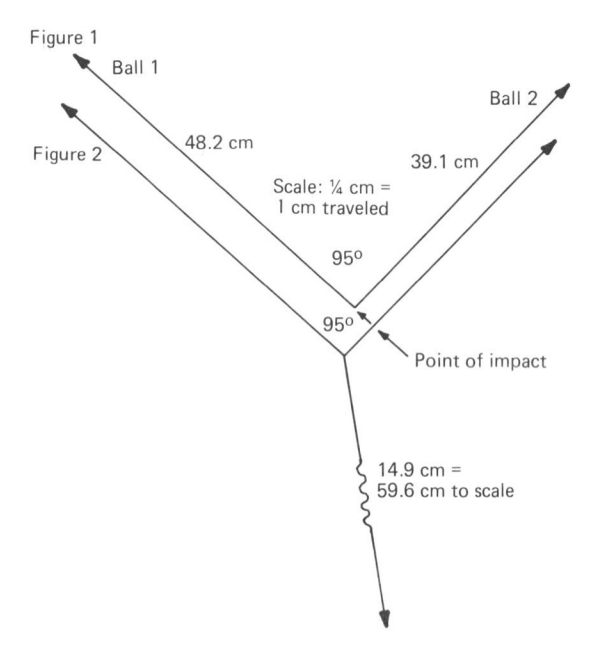

Ball 1

Figure 2

48.2 cm

Ball 2

39.1 cm

Scale: ¼ cm = 1 cm traveled

95°

95°

Point of impact

14.9 cm = 59.6 cm to scale

Figure 3

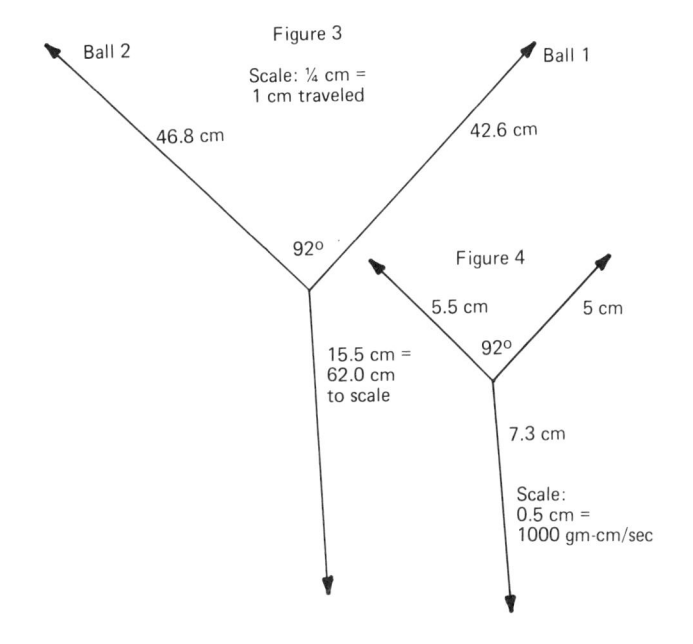

Scale: ¼ cm = 1 cm traveled

Ball 2 — 46.8 cm

Ball 1 — 42.6 cm

92°

15.5 cm = 62.0 cm to scale

Figure 4

5.5 cm

5 cm

92°

7.3 cm

Scale: 0.5 cm = 1000 gm-cm/sec

Therefore: 5.5 cm = 10,500 gm-cm/sec
5.0 cm = 10,000 gm-cm/sec
7.3 cm = 15,645 gm-cm/sec

Experiment 4—A Collision in Two Dimensions

The "Collision in Two Dimensions" does not show a laboratory here. But the idea is closely related to Experiment 3, where the reader understands the potential step for another good collision laboratory item. You may want to work with ball collisions, for which the classes teacher will give extra credit.

Individual laboratory steps add up to excellent personal education. Related to this, the PSSC Physics book (if available at your school) shows Lab #28, on this subject.

Experiment 5—Gun Muzzle Velocity

This experiment is for everyone. Work in pairs, or threes; for data, write up and analyze your own work. This is a *major* laboratory experiment in terms of learning and grades, which will reflect your complete knowledge on the principles of Newtonian physics.

Instructions

"Shoot" a known mass (ball) forward and determine where it lands—how far away from the starting point is it?

Pull back the "gun" at constant distance (exactly 90° from the vertical) and hit the ball of known mass. Then, determine the distance traveled—constantly—as it hits the floor.

Then repeat this with at least three other, balls. In each case, take a ball of *unknown* mass and through this method determine its mass.

Related to these scientific calculations about distance traveled, write a scientific, complete essay or report on your laboratory observations. With this, include as much and many of the physical Newtonian principles as possible:

Mass, weight, time, distance
Speed, velocity, acceleration, vectors
Force, momentum, inertia

Gravitational force
Work, energy—potential and kinetic

With this write-up, also show as many graphic relationships (and vectorial ones) as possible.

Experiment 6—The Spring: Shooting a Ball Forward

For this setup, use one ball. The experiment involves vector consideration. The question is; At what angle (shot forward and upward) does the ball go farthest forward? Again, this experiment will involve *many* Newtonian principles. You will observe the demonstration and then go on your own to work out details.

Experiment 7—Analysis of Newtonian Physics with the Whirligig

For this experiment, there is an illustration of the Whirligig. It shows a whirligig item that can contan varying masses on the "outside" points. Associated with the round center that whirls is a rope which pulls down a variety of other masses. Also, the falling mass goes back up at a given distance. One can clearly see the complexity and *fun* of these steps—and all the scientific potentials illustrated. The location is up in the air—about ten feet—for the mass to fall significantly. The experiment shows force, friction and mass relationships. Complete analysis adds up to personal knowledge of most all aspects of mechanics.

This is a laboratory item that can be made by students for their own personal education and even extra learning steps that add up to higher learning points: With this experiment, you will have very few basic instructions. You *should* be able to go through the process and come up with ideas of velocity, momentum, force, acceleration, potential energy and kinetic energy, time and mass, relationships, and so forth.

The whirligig goes around, pulling by a dropping mass. The mass can be varied on the ends of the whirligig. Another item is a stopwatch. This experiment may *seem* simple. It is your responsibility to "dig out" as many observational directions and as much data as you can.

Experiment 8 – The Bicycle Experiment

This experiment is about the physics of old-time human motion—energy of a boy plus a bicycle; the mechanics of motion.

Equipment:

Stopwatches to analyze time
A 100-meter road (less distance can also be used)
One-speed bicycle, or a bicycle with more speed-steps if others are not available.
Student bicycle rider and other students as timers

Measurements:

Weight (Mass) of student, in kilograms
Mass of the bicycle
Total Mass—the bicycle plus rider

Distances Traveled:

Zero meters, 10 meters, 20 meters, 30 meters, 40-100 meters (noted as distances of 1, 2, 3, 4, 5)

Time to reach each ten-meter distance

Ten-meter distances: 0, 1, 2, 3 ... 11
Time from 0 to 1, 1 to 2, 2 to 3, 3 to 4 ... 10 to 11
Total time—the distance from 0 meters to 100 meters; distance 0 to 11
Then, total measurements equal: time, distance, and mass

Physical Calculations:

Determine from the Relationships of mass, time, and distance, the speed at each 10-meter interval between each interval and find the average for total distance. Here speed equals velocity

Acceleration—at and between each 10-meter interval, for the total distance. This represents the changing velocity. Then find the:

Average acceleration, and greatest acceleration, least acceleration
Momentum, and Inertia
Force if possible, at each 10-meter distance, for the total weight, and gravitational force
Potential energy
Kinetic energy

These Calculations can then be written up as a graphical analysis.

This experiment represents the BASICS of all the ideas related to Newtonian physics education. You will be required to write a *complete* and *significant* essay/lab report on all aspects. All of the data should be done before the snow time! It can give you two or three class days to work on this. Also your essay should show data tables and graphs, of as much as possible.

For each class, at least five (more, if possible) students should ride the bike, and then, take data on each and, for the students from all classes also calculate the amount of potential energy per individual—the "best" and the "least."

Experiment 9—Falling Bodies and the Swinging Pendulum

This experiment, along with all the steps noted here, may be a laboratory test item. Be sure to read and analyze all the steps carefully.

This is simple harmonic motion. We will relate the length of the pendulum to time of swing (varying lengths) and then calculate the other related factors. Mass of Ball—assume as *one gram*.

1. Record the following data:

Pendulum Length (meters)	Time of 10 swings	Time of 1 swing	Pendulum L, Meters	Time of 10 swings	Time of 1 swing
1.6	_____ ____		0.8	_____ ____	
1.4	_____ ____		0.6	_____ ____	
1.2	_____ ____		0.4	_____ ____	
1.0	_____ ____		0.3	_____ ____	

From the above data only, try to come up with your analysis of what is constant for these relationships. (This analysis does not have to be "right"—only a *good scientific view* of Newtonian interrelationships.)

2. What is the relationship between length and time of swing? (Again —a *preliminary* analysis.)

3. Use the attached pendulum diagram to calculate the distance that ball moves to the center of the swing in time. Measure distance and calculate time (1 m equals 20 cm). (B=¼ total swing.) See diagram, next page.

A. Length of Pendulum (meters)	B. Distance that Ball Moves to center of swing	C. Time X to Y (seconds)	A.	B.	C.
1.6	_____	_____	0.8	____	____
1.4	_____	_____	0.6	____	____
1.2	_____	_____	0.4	____	____
1.0	_____	_____	0.3	____	____

4. From the foregoing, calculate the acceleration (see "*Rational Approach*" chapter for formula) for each length. Show your own work.

1 = 1.6 m,   2 = 1.4 m,   8 = 0.3 m, and so forth.

1.   5.
2.   6.
3.   7.
4.   8.

5. For the pendulum swinging, it goes *faster* as it becomes *shorter*. How does this relate to gravitational pull (force)? Is it or should it be constant? Relate this to your other calculations and show the mathematical development.
6. Formula:

$$T = 2\pi\sqrt{\frac{L}{g}} \quad \text{so} \quad T^2 = \frac{4\pi^2 L}{g}$$

L = length
g = gravity
T = time

These formulas relate mathematically to your laboratory calculations and observations. Use them for items below—showing all work—and see how they relate to your experimental analysis.

2.   4.
6.   8.

7. What can you generally say about the motion of a pendulum? Can you relate its motion to gravity? How? Show this vectorially. What is the force?
8. We will take another look at this ball pendulum in relation to a freely dropping ball. Possibly, more analytical questions will follow.

Diagram: Swinging Pendulum

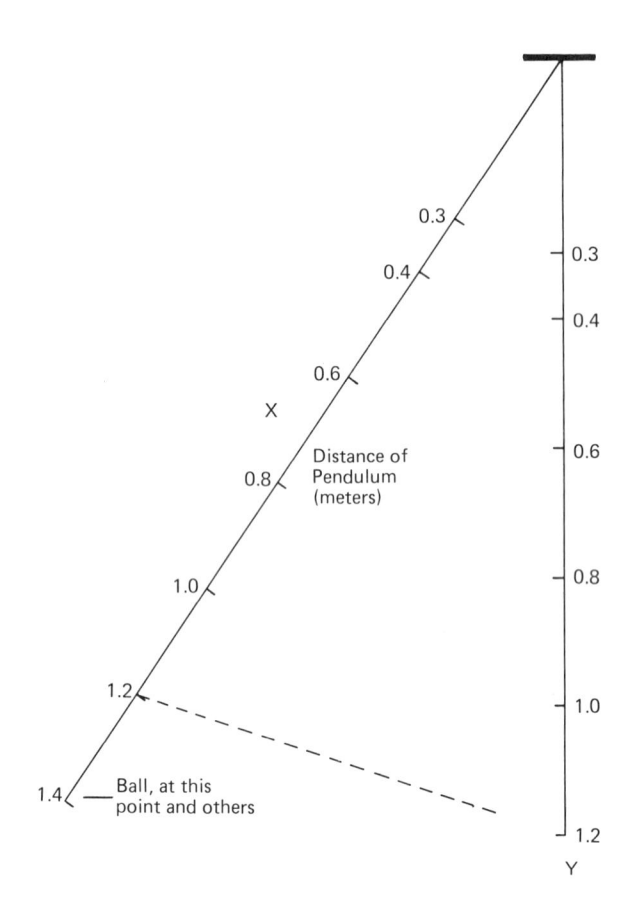

X

0.3
0.4
0.6
0.8
1.0
1.2
1.4

Distance of Pendulum (meters)

Ball, at this point and others

0.3
0.4
0.6
0.8
1.0
1.2

Y

8

Review Questions on Newtonian Mechanics

1. Relating kinetic and potential energy.

    A mass of 10 kg is placed at a height of 122.5 m above the earth. Fill in the Data Table below with all proper figures and show E_k and E_p for the system at every half-second level.

    Make a graph which shows ΔE_k and ΔE_p for this system as indicated below. On another paper, show your method of calculations stepwise.

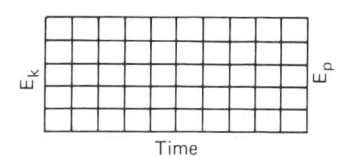

    Data Table:

Distance traveled (meters)	Time Point (sec)	Acceleration (m/sec^2)	Velocity (m/sec)	Momentum	Acting Force	E_p	E_k
——	0	——	——	——	——	——	——
——	0.5	——	——	——	——	——	——
——	——	——	——	——	——	——	——
——	——	——	——	——	——	——	——

This Data Table should have twelve lines, as noted above here. Make your own set of these items.

2. Compare the kinetic energy of two objects, A and B, identical in every respect except one. Assume that the single difference is:

   a. A has twice the velocity of B.
   b. A moves north, while B moves south.
   c. A moves in a circle, while B moves in a straight line.
   d. A is a projectile falling vertically downward; B is a projectile moving vertically upward at the same speed.
   e. A consists of two separate pieces attached by a light string, each equal in mass to the mass of B.

3. A mass of 4.0 kg sliding with a speed of 3.0 m/sec on a frictionless horizontal table, collides with a queer type of spring bumper. The bumper exerts a constant force of 120 newtons on the mass as it moves in (compressing the spring) and the same force on the way out, until the spring is back where it was.

   a. Is this an elastic collision?
   b. What is the kinetic energy at the beginning of the interaction?
   c. How much is the spring compressed?
   d. What is the ratio of kinetic energy to potential energy when the spring has been compressed 10 cm?

4. Two masses, 1 = 1.0 kg and 2 = 2.0 kg, are pushed together despite an elastic interaction force, until their potential energy is 27 joules greater than it is when they are beyond the distance at which they interact. If they let go, what will be the final kinetic energy of each?

5. A force of 30 newtons accelerates a 2.0 kg object from rest for a distance of 3.0 meters along a level, frictionless surface. The force then changes to 15 newtons and acts for an additional 2.0 meters.

   a. What is the final kinetic energy of the object?
   b. How fast is it moving?

6. An automobile of mass 1000 kg is moving at 100 km/hour. This speed is read on the speedometer.

   a. What is its kinetic energy?
   b. How much work was done to provide this kinetic energy?
   c. Can you determine what force acted on the car to provide this kinetic energy? Determine the distance through which this force acted.

7. A man pulls with a string on a 20.0-kg mass that was initially at rest on the floor. He exerts a force of 20.0 newtons horizontally, and the mass moves through 8.0 meters. The mass then has a velocity of 3.00 m/sec.

   a. What is its final kinetic energy?
   b. How much energy has been transferred from the man?
   c. How do you explain the difference in your answers to a. and b.?

8. A 100 gram light bulb dropped from a high tower reaches a velocity of 20 m/sec after falling 100 m. About how much energy has been transferred to the air?

9. A force of 3.0 newtons and a force of 4.0 newtons are applied simultaneously to a 4.0 Kg mass that is initially at rest. The two forces act at an angle of 90 degrees with each other for a 2.0 seconds.

   a. What is the net force?
   b. Use the net force to calculate the total work done.
   c. What is the work done by each separate force? How does the sum of these works compare with the answer to b.?

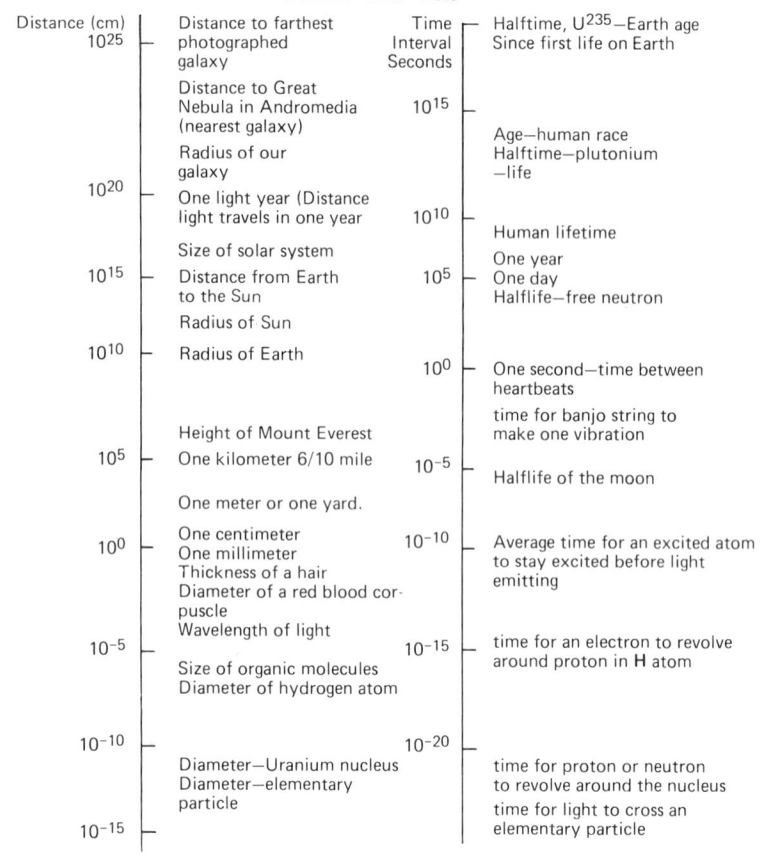

Magnitude of Distances

Distance—Time—Mass

Distance (cm)		Time Interval Seconds	
10^{25}	Distance to farthest photographed galaxy		Halftime, U^{235}—Earth age Since first life on Earth
	Distance to Great Nebula in Andromedia (nearest galaxy)	10^{15}	
	Radius of our galaxy		Age—human race Halftime—plutonium —life
10^{20}	One light year (Distance light travels in one year)	10^{10}	
	Size of solar system		Human lifetime One year
10^{15}	Distance from Earth to the Sun	10^{5}	One day Halflife—free neutron
	Radius of Sun		
10^{10}	Radius of Earth		
		10^{0}	One second—time between heartbeats
	Height of Mount Everest		time for banjo string to make one vibration
10^{5}	One kilometer 6/10 mile	10^{-5}	
	One meter or one yard.		Halflife of the moon
10^{0}	One centimeter One millimeter Thickness of a hair Diameter of a red blood corpuscle	10^{-10}	Average time for an excited atom to stay excited before light emitting
	Wavelength of light		
10^{-5}		10^{-15}	time for an electron to revolve around proton in H atom
	Size of organic molecules Diameter of hydrogen atom		
10^{-10}		10^{-20}	
	Diameter—Uranium nucleus Diameter—elementary particle		time for proton or neutron to revolve around the nucleus
10^{-15}			time for light to cross an elementary particle

Magnitude of Masses

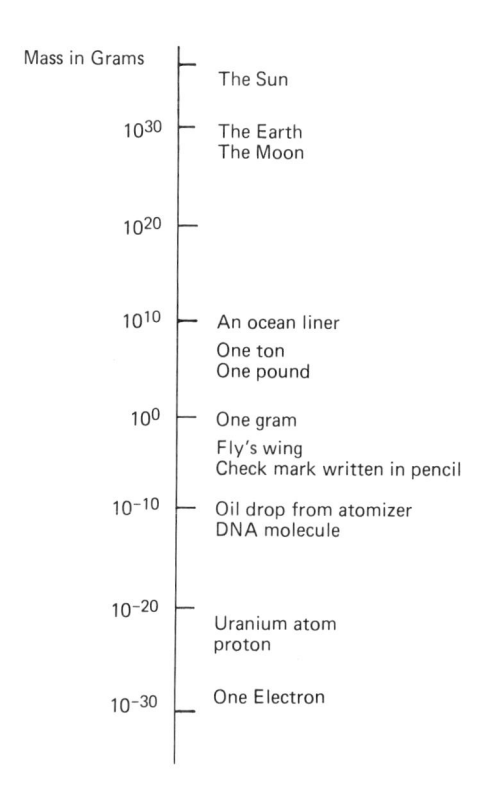

Mass in Grams

- — The Sun
- 10^{30} — The Earth
-         The Moon
- 10^{20} —
- 10^{10} — An ocean liner
-         One ton
-         One pound
- 10^{0} — One gram
-         Fly's wing
-         Check mark written in pencil
- 10^{-10} — Oil drop from atomizer
-         DNA molecule
- 10^{-20} —
-         Uranium atom
-         proton
- 10^{-30} — One Electron

9

Newtonian Problems (Homework Items)

This is a list of potential problems you will do as you progress in steps and understanding. Some will be handed in for credit; others will be discussed for learning development. These problems are from books other than your text; some problems may come from this also. We may also find other good examples as we progress. You'll probably get weekly assignments on these items.

I. A number of questions in your handout book—the first seventeen pages, generally, such as: Where does *mass* come in?

II. Titles (subject areas) from other books—and the questions.

The Description of Motion

1. In the first event of the 1974 World Cross Country Cup series in skiing, Dieter Klause of East Germany took first place, covering the 30-km course in one hour, twenty-eight minutes, and forty seconds. What was his average speed in meters per second?
2. A man drives a car 40 miles in one hour, and then 30 miles in the next half hour in the same direction: (a) How far has he traveled? (b) What is his average speed for the whole trip?
3. The world record in the 20,000-meter race is held by Gaston Roelants of Belgium, with a time of fifty-eight minutes, and six seconds. What was his average speed in meters per second for the race?
4. The total distance traveled on an automobile trip is 510 miles. If the total driving time is seventeen hours, calculate the average speed.

5. A falling rock on earth achieves a velocity of 49.0 m/second. How long has it been falling?

Force, Mass, and Newton's Second Law

6. A drag racer accelerates from a standing start to 60 m/second in twelve seconds. (a) What is the average acceleration? (b) If the mass of the car is 1,000 kg, what force was applied to it to give this acceleration?
7. A golf ball of mass 0.06 kg driven by a golf club acquires a speed of 80 m/second during the impact. The impact lasts 2×10^{-4} second. Assume that a constant force acts on the ball. (a) What is the acceleration of the ball? (b) What force acts on the ball?
8. An object weighs 450 pounds on the moon. What does it weigh on earth?
9. A force of 6 newtons acts on a 10-kg mass for 10 seconds. (a) Find the acceleration. (b) How far does the object move in these 10 seconds?
10. A bullet of mass 10 grams strikes a tree and stops in 0.1 meters. Its velocity was 300 m/second. Find the average force on the bullet.
11. A car of mass 2,000 kg is observed to stop from a speed of 30 m/second in a distance of 90 meters. (a) Calculate the braking force, assuming that it is constant. (b) Calculate the acceleration. (c) What is the weight of the car?
12. A car of mass 1,200 kg accelerates from rest to a speed of 30 m/second over a distance of 400 m. (a) What is the acceleration of the car? (b) How long does it take the car to cover 400 m? (c) What force acts on the car?
13. In an effort to determine the acceleration due to gravity on a strange planet, a ball is dropped a distance of 200 meters. It is found that it takes five seconds for the ball to drop this distance. The ball has a mass of 5 kg. (a) What is the acceleration due to gravity on this planet? (b) What is the velocity of the ball as it passes the point 200 meters below the point at which it was released? (c) What does the ball weigh on this planet? (d) What would the ball weigh on the Earth?

Circular Motion and Gravity

14. Why are there seasons?
15. As a planet moves about the sun, its orbital velocity changes so that equal areas are swept out in equal times by a line joining it and the sun. In what orbital position does the planet have the largest orbital velocity? Where does it have the smallest orbital velocity?
16. A boy swings a ball on a string, around his head whose mass is 0.1 kg. The string is 3 meters long. If the force required to break the string is 20 newtons, how fast will the ball be going when the string breaks?
17. Calculate the inward acceleration for a car that is traveling on a curve of radius 300 meters at a velocity of 40 m/second. If the car has a mass of 2,000 kg, how much force is needed to keep it in the road?
18. Calculate the gravitational force between two objects whose masses are 2 kg that are 10 meters apart.
19. (a) Calculate the velocity of a satellite in orbit above the Earth at a distance of 5×10^7 m (about 30,000 miles) from the center of the Earth. (b) Calculate the period of such a satellite.
20. The Earth rotates on its axis such that a point on the equator makes one complete rotation in one day. The average radius of the earth is 6.38×10^6 m. (a) How far does a person on the equator travel in one day because of this rotation of the Earth on its axis? (b) What is the velocity of the person on the equator due to this rotation? (c) What is the acceleration of a person on the equator due to this rotation?

Newton's Third Law and Momentum

21. A baseball which has a mass of about 250 grams is thrown with a speed of 40 m/sec (about 90 miles/hour). What is its momentum?
22. In the experiment of two cars on a frictionless track, tied together with a compressed spring between them, the string is gently cut and the cars move apart due to force of the spring. This experi-

ment is performed using two masses. Mass 1 is 1 kg, and mass 2 is unknown. The velocities after separation are: mass 1–5 m/second, and for mass 2–3 m/sec. What is the mass of 2?

23. (a) How fast would a car have to go in order to bring a truck that weighs ten times as much to a dead stop in a head-on collision if the truck is going 20 miles/hour and both are coasting? (b) What is the momentum of the system (car plus truck) before the collision? (c) What is the momentum of the car/truck system after the collision?

24. Two automobiles collide in a completely inelastic collision. Initially car B, whose mass is 1,500 kg, was at rest. Car A, which has a mass of 2,000 kg, has been moving 25 m/second before the collision. Find the velocity of the wreckage.

25. A cannon on wheels that has a mass of 200 kg, shoots a shell of mass 5 kg. If the shell leaves the muzzle with a horizontal velocity of 500 m/second, what is the backward velocity of the cannon?

26. A golf ball whose mass is 0.2 kg is struck and given a velocity of 60 m/second. If the impact between club and ball lasts 10^{-3} second, what is the average force on the ball during this time?

Work and Energy

27. (a) Calculate the potential energy change in lifting a 10-pound weight a distance of seven feet. (b) Calculate the potential energy change in lifting this same object seven feet on the moon.

28. A 10-kg mass passes a window falling at a rate of 50 m/second. (a) Neglecting air resistance, from what height did this object fall? (b) What was the potential energy of this object before it became to fall?

29. Using the law of conservation of energy, find how a ball thrown upward with a velocity of 5 m/second will rise.

30. A girl throws a 0.2-kg ball a distance of 6 meters straight up in the air. (a) What is the potential energy of the ball at its highest point, relative to the point of release? (b) What was the kinetic energy of the ball as it left the girl's hand? (c) How much work does the girl do in throwing the ball? (d) If the girl's arm muscle

contracted a distance of 0.5 m while throwing the ball, what was the average force exerted by the muscle?

31. A block weighing 6 pounds slides down a frictionless chute and then strikes a stationary 10-pound block on a horizontal table. After the collision, the blocks stick together. (a) What is the velocity of the first block when it reaches the bottom of the chute? (b) After the collision what is the velocity of the two blocks?

32. A body of mass 5 kg coasts up an incline. As it starts up the incline, its speed is observed to be 4 m/second. At the top of the incline it comes momentarily to rest and then slides back down the incline. When it reaches the bottom of the incline, its speed is 3 m/second. (a) How much mechanical energy is converted to thermal energy during the roundtrip? (b) If one-half of this energy conversion occurs during the trip up, how high above the lowest point on the incline does the block rise?

33. An automobile of mass 2,000 kg collides with another of mass 1,500 kg. Before the collision the first car has a velocity of 30 m/second, and the second velocity of 10 m/second. If, after the collision, both cars have a velocity of 20 m/second, how much kinetic energy is lost in the collision?

34. Two cars collide head-on and stick together. Immediately after the collision, the wreckage is observed to be at rest. The first car weighs 3,200 pounds and before the collision, it had been traveling north at 60 miles/hour. The second car was heading south before the collision at a speed of 50 miles/hour. (a) What is the weight of the second car? (b) What was the total kinetic energy after the collision? (c) What is the total kinetic energy before the collision? (d) How much kinetic energy is lost and where might it possibly have gone?

10

Vectors—Items to Analyze

We have considered the motion of a ball in a straight line only. Δd tells us how far the ball traveled. For any path other than a straight line, direction must be added to the magnitude (linear extent) of the displacement number to correctly and completely describe motion. Any quantity that *requires both magnitude and direction* in its description is a *vector*. Figure 1 is a map of the Midwest with Chicago, Kansas City, and Saint Louis identified. Grid (graph) squares are twenty kilometers on an edge. Compass directions are also shown. Assume that you drive from St. Louis to Chicago. (Note: $\Delta d = \Delta x$, where x = position; so Δx is change in position)

1-6 Calculate the following:

1. Magnitude in kilometers (km) for the trip.
2. Approximate compass direction of the travel.
3. Are both a direction and a distance magnitude necessary to describe a trip from St. Louis to Chicago?
4. Describe this trip as a *vector* quantity.
5. In what way does the vector from Chicago to St. Louis differ from the vector from St. Louis to Chicago?
6. If you traveled from St. Louis to Chicago by way of Kansas City, would this change the vector from St. Louis to Chicago? If you only wanted to get to Chicago, how far would you have effectively traveled?

The idea of a *vector sum* or *resultant* will be clarified. The vector we have been analyzing is a displacement vector. The displacement from one point (city) to another does not change, no matter what route one takes to get there. We will also analyze velocity and force vectors.

A quantity that does not require direction in its description is called a *scalar*.

7. Is "390 km" a scalar or a vector quantity?
8. Is "390 km, northwest" a scalar or a vector quantity?

Referring to the figure, consider a six-hour trip from St. Louis to Kansas City.

9. What is the average velocity for this trip?
10. Does your answer completely describe the motion?

If you did not include direction in your answer to question 9, you did *not* completely describe the velocity. Velocity includes both magnitude and direction of motion.

11. Is velocity a vector or a scalar quantity?
12. A velocity term includes _____ and _____ and _____.
13. "Speed," a scalar quantity, includes _____ and _____.
14. Scalars are *undirected* quantities. Which of the following terms are scalars and which are vectors?

    a. The number of students in the classroom.
    b. A temperature of 10° C.
    c. A speed of 10 m/sec.
    d. A velocity of 10 m/sec east.
    e. A ball thrown straight up at an initial speed of 5 m/sec.
    f. Your mass.
    g. Your weight.

An ocean vessel is heading east at 4 m/sec. Its velocity is represented by the arrow below.

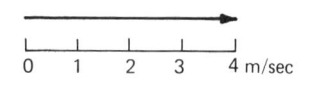

```
         ──────────────▶
    └────┴────┴────┴────┘
    0    1    2    3    4 m/sec
```

A passenger walks north across the ship's deck at 3 m/sec. His velocity is represented by the arrow following.

```
 3 ┌  ▲
   │  │
m/sec 2│  │
   │  │
 1 ├  │
   │  │
 0 └  │
```

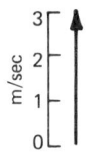

15. What is the approximate compass direction of the passenger's motion?
16. How would you determine the magnitude (size) of his velocity?

Figure

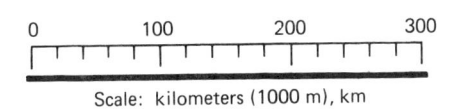

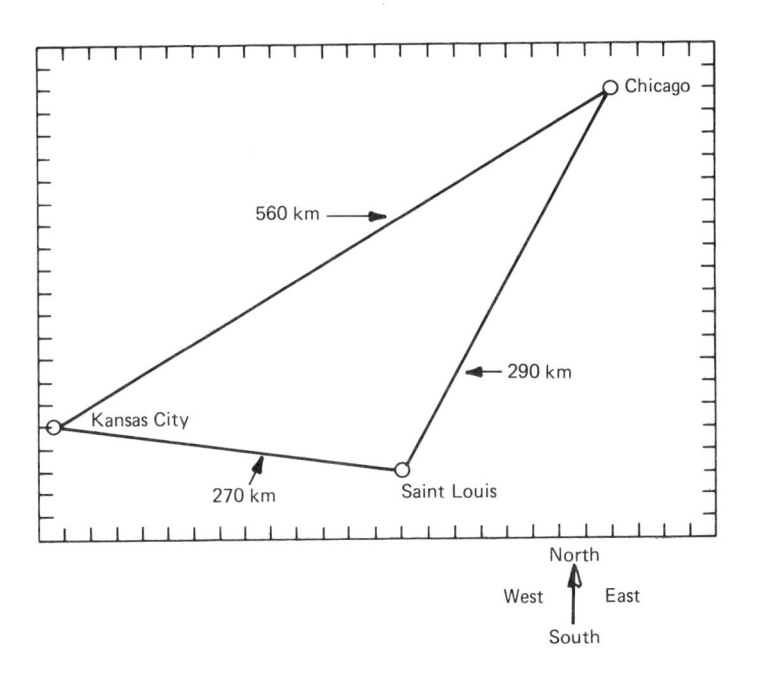

Scale: kilometers (1000 m), km

- Chicago
- 560 km
- 290 km
- Kansas City
- 270 km
- Saint Louis

North
West — East
South

Questions 15 and 16 are problems in *vector addition*. The sum of two vectors cannot generally be obtained by adding magnitudes. (The magnitude of the velocity for our passenger is not 7 m/sec.) Vectors are described by arrows, which must *always* be drawn to scale in the proper direction. A vector sum is obtained by placing the tail of either vector arrow at the head of the other vector arrow. The diagram below shows vectorial sum of the 4 m/sec east plus 3 m/sec north. R is the symbol for a resultant sum of the two vectors.

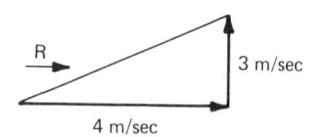

Displacement and velocity are both vector quantities.

The following problems require vector addition.

17. An airplane is heading north at a forward speed of 180 km/hr. It is displaced sideways by a wind blowing toward the east at 30 km/hr.

    a. What is the resultant velocity of the plane? Show a vector diagram for determining this velocity.
    b. Is the velocity constant?
    c. What is the approximate compass direction of flight?
    d. The displacement is not constant. Calculate the displacement after two hours. Describe it fully.

18. A man can row his boat at a speed of 4 km/hr. He heads west directly across a river that is flowing south at a velocity of 4 km/hr.

    a. Draw a vector diagram to indicate the boat's resultant speed and direction.
    b. The river is one kilometer wide. How far will the boat travel before it hits the opposite side of the river?
    c. How much time will it take him to cross the river?

d. How far down the river will he have traveled when he reaches the opposite bank? (This question will require a new displacement vector diagram.)

19. What would your total displacement (resultant) be if you made a trip 1 km east, then 2 km north, then 1 km west, and finally 2 km south?

Look again at our definition of average velocity. Velocity is a vector quantity. When motion is restricted to a straight line, as we have done in the laboratory, direction is possible only in a forward and backward direction. This is specified by the symbols, (+) (forward or to the right) and (−) (backward or to the left).

20. A ball moving along a straight line starts at the 3 meter mark and moves to the 7 meter mark. What is its Δd?
21. A ball is given a displacement of +10 m, followed by a displacement of −3 m. What is the Δd (resultant displacement)?
22. An object moving in a straight line passes the 4.5 meter mark when a clock reads 2 sec. It passes the 10 meter mark when the clock reads 6 sec. What is its average velocity?
23. An object moving in a straight line passes the 10 meter mark when the clock reads 2 sec. It passes the 4.5 meter mark when the clock reads 6 sec. What is its average velocity?

Data in the table below was taken in the laboratory in which a ball moved along a straight line. The following questions relate to the data.

Table

Time (sec.)	Distance (m)
0	0
1	2
2	4
3	6
4	6
5	6
6	3
7	0

24. Draw a graph of this data.
25. What is the average velocity for the first three seconds?
26. What is the average velocity during the fourth and fifth second intervals? (Hint: When did the fourth second interval begin?)
27. What is the average velocity during the last two seconds?
28. Write a paragraph describing the motion of the ball.

11

Items on Momentum

In a system that consists of bodies on which no outside forces are acting, the total momentum equals _____ (finish the sentence).

This is the Law of *"GGG."*

Related to the law above is our two-ball (billiard balls) collision in the lab experiment. Diagram and explain. For the illustration of the little cars on different surfaces, going at equal distances explain and relate.

Transfer of Momentum

Illustration: balls on strings. What can you say about the associated masses, acceleration, velocity, force, and momentum?

A firecracker explodes into two pieces, number one of which is much larger than number two. Which piece will have the greatest velocity? Which will have the greater momentum?

What is *impulse*, and how is it related to momentum?

A 1500-kilogram car has stalled on a level road. The coefficient of rolling friction between the tires and road is 0.03. How long would one have to push horizontally on the car with a force of 1.3×10^3 newtons to impact to it a speed of 1 m/sec? What assumptions have you made?

Relate the sum of momentum for two isolated systems, sum M_a is proportional in what respect to sum M_b? Consider an isolated system, in which the number N bodies act upon one another. Discuss this. Any two interacting bodies with a change in **M** (momentum) are what in magnitude, and direction?

Relate these ideas to a change in momentum, (M) for teams A and B when there are eleven football players interacting. Make an equation that shows this. Try to think of other "real-life" phenomena that relate to physical principles.

A container of gas molecules at 20 degrees Celsius. The number of interacting molecules is in the billions. Each of the colliding molecules continuously undergoes velocity and vector changes; discuss and explain this.

A ball of mass ____ inelastically collides with a ball of mass ____ and initial velocity of ____ toward the (direction). What must be the initial velocity of the first ball so that the final velocity of the pair will be ____?

This is a sample question that may be given on a test.

Explain the ideas of *Conservation of Momentum*.

Discuss the validity of this statement: If you jump upward, the Earth will recoil with an initial momentum that is equal in magnitude but opposite in direction to your initial momentum. During each moment that you are in the air, your momentum, and that of the Earth will be equal in magnitude and opposite in direction.

The net force acting on a body is equal to what in terms of momentum?

Is momentum vectorial or not?

12

Energy Items and Ideas

Two items can have (must have, if related) equal momentum, but not equal kinetic energy:

$$m^1 v^1 = m^2 v^2$$

$$\text{kg} \times 2 \text{ m/sec} = 2 \text{ kg} \times 1 \text{ m/sec}$$

1. But—What are the relative kinetic energies of 1 compared to 2? Work this out and explain, using logic.
2. Related to this, and associated with Section 6.6:
   What must the velocity relationship be for momentum to equal K.E.?

   The one-kilogram mass has a velocity of 2 m/sec

As a general mathematical relationship, for a "double mass" of one to equal the "single mass" of the other in terms of kinetic energy, work out the relationship in an equation.

$$KE^{\text{double mass}} = KE^{\text{single mass}}$$

3. A specific example:
   Mass A = 4 kg, and has a velocity of 2 m/sec
   Mass B = 2 kg, so its velocity must equal—Work this out.
4. Note demonstration one, a fulcrum.
   Mass = one kilogram. We can read F in newtons at the end of the fulcrum. Show how to calculate F where the mass is, and explain the relationship.
5. If your mass is 90 kg, what lever setup would you need to lift a 900 kg mass?

6. If you know the mass and acceleration of an object, can you figure out the KE? If not, what additional information would give us this quantity? Work this out mathematically. Answer: one must know _____.
7. Units of work, energy, and power
The unit of work is the joule. One joule = 1 nt x 1 meter, or = KE
   Joule = J (KE *is* work *done*.) Go through example 6.2. And Fd = m a d, equals *good* for the relationships discussed.

Ideas on Power

A light bulb reads 60 W, 130 V. If it costs 10 cents per kilowatt hour to use the light, what is the cost per 10 hours?

8. This is associated with section 6.3, on potential energy. On the frictionless air track equipment in the lab, make it level, and bounce a "car" off one end, time it. Calculate: PE and KE. Relate to **F**, acceleration, mass, and velocity.
9. Note the equations. A mass of 10 kg accelerates 10 m/sec^2. Associated with this, what happens to the time with the kinetic energy? As time doubles (x2, each step), joules increase by how much? Show calculations for the first through eighth seconds.
10. PE = energy of position, or stored energy. Raising something gives it PE; Dropping yields KE. **PE = mgd** = ½mv^2 = **KE**.
    PE = sum of work done against a force, so it is *stored* for future release. Relate this to learning, where PL equals potential learning. Calculations:

    PL = sum of KL, kinetic learning; BC = brain capacity
    KL is ½ x brain capacity x (learning speed)^2 = ½ BC x LS^2 or W = F x d = mad. So, BC x LS x t

11. As mass doubles, what happens to KE? As velocity doubles, what happens to KE?
    If mass doubles, and velocity halves, what happens to KE?
12. What physical facts do you have to consider to catch a 5-Kg trout with a line that will *break* when it holds more than 5 kg, vertically?
13. View and analyze the principles of mechanics as applied to human anatomy.

13

Newtonian Review

Terms

inertial mass	fulcrum	slopes
momentum	angular momentum	average velocity
Newton	Newton's Laws	tangent
force of gravity	circular motion	9.8 nt.
distance	Kepler's Laws	9.8 m/sec^2
time	height	kinetic energy
mass	total work done	potential energy
speed	friction	work
velocity	elastic	energy of position
weight	inelastic	energy of motion
inertia	watt	joule
slope	unit of force	conservation
second	law of inertia	of energy
power	energy	centripetal force
meter	velocity	centrifugal force
kilogram	instantaneous	weightlessness
gram	velocity	kilometer
gravity	increase	millimeter
incline	decrease	elapsed time

Symbols

d, t, v, F, V_f, V_i, E, PE, KE, a, E, G, W, g, kg, W, V^2
Δv, m, E_p, E_k, V_{av}, d_f, d_i, Δd, Δt, t_f, t_i, nt, Δ, h

Equation Relationships

v/t,　　d/t,　　a/t,　　d/t², 　　md/t², 　　md²/t², 　　Fd,　　mad,　　(d/t)²,

mv²,　　gt,　　½at²,　　mad,　　v²,　　mv²,　　$V_f/2$,　　½mv²,　　v²/r,　　v/Δt,

V_f-V_i,　　(m/sec)$_f$-(m/sec)$_i$,　　m/sec²,　　kg-m/sec/sec,　　f/m,　　mΔv,

mv,　　mg,　　ΔKE,　　ΔPE,　　F×d,　　kg-m²/sec²,　　m-v²/r

Problems

Write seven equations for KE. Write nine equations for acceleration.

One problem for the Newtonian test may be to write down all the equations that you can. For example if you have _____ and _____ information, what can you calculate about the situation? (*i.e.,* momentum, mass, distance, KE, work, velocity, acceleration and so forth). Show the equation(s) needed.

14

Quizzes and Tests

Sample Test

Straight Line Motion Studies

This is a sample test. Fill in your own sample numbers for practice. (Work with friends, if you like, but be sure that you know how to do it.)

An object starting from rest has a constant acceleration rate of ___ m/sec^2 for ___ seconds. At the end of this time, it has zero acceleration for seconds. After this time period, it comes to rest (zero velocity) in ___ seconds.

Calculate or describe the following:

1. Average velocity for the first ___ seconds.
2. Average velocity for each second time interval from 0 to B.
3. Distance traveled in the first ___ seconds.
4. "Final" velocity at the end of each second from 0 to B.
5. Acceleration from times B to C.
6. Distance traveled in time B to C.
7. Distance traveled in time 0 to C.
8. The average acceleration rate (assumed to be constant) in the D-second time interval.
9. Average velocity in the D-second time interval.
10. Distance traveled in the D-second time interval.
11. Average velocity for the total time interval.
12. Total distance traveled.
13. Average acceleration for the total time interval.

14. Draw a distance time graph for the total time interval.
15. Draw a velocity time graph for the total time interval.
16. Draw an acceleration time graph for the total time interval.

Using a graph of velocity versus time, in which the velocity

First, increases   Second, becomes constant
Third, increases more   Fourth, decreases to zero
First = A to B   Second = B to C   Third = C to D   Fourth = D to E.
With this graph, make a graph of distance time for points A, B, C, and D
With this graph, make another graph of acceleration time for points A, B, C, D, and E

Football Quiz—Related Items

These items are associated with the diagram on the next page, also with different speed and beginning area.

You are running north to the point of touchdown on the metric football field. You start at a point five meters east (from west) and 10 meters from the south end. Your constant speed forward towards the north is 5 m/sec. When you hit the 20-meter line, another competing player, is pushing you directly east at a constant speed (for his force) at 2 m/sec. (Assume that no other competitor comes by.) Answer the following questions with mathematical explanations.

1. On the diagram (make one of your own) show your motion from Point X (the beginning step) toward the north *with* the push from the competitor. Make a *full line* representation for this and relate the other questions to your analysis.

You may want to work out the other questions first. Calculate the information and write it down.

2. Where did you wind up on the field; either touchdown or going off bounds before touchdown?
3. Do you make the touchdown? Explain.
4. Through a diagram and mathematical calculation, show your velocity.

Football Quiz

You are running north, straight down a metric football field along one side (west) and assume that you are exactly along the west sideline at 4 meters per second, constant speed. When you hit the 20-meter south line, another player is pushing you directly east at a constant speed of 2 meters/sec. (Assume that no other player comes by.) Answer the following questions with mathematical explanations. (Answer on the back of this paper.)

1. Do (did) you make the touchdown? Explain.
2. Where did you wind up on the field (either touchdown or going off bound)?
3. How long did it take you to go to the point mentioned in question #2?
4. Show your velocity by diagram and mathematical calculation.

Another crazy sports quiz that you can write involves a kick after touchdown: distance kicked, force kicked, angle, mass of ball, height of bar, and so forth. Did you make the kick?

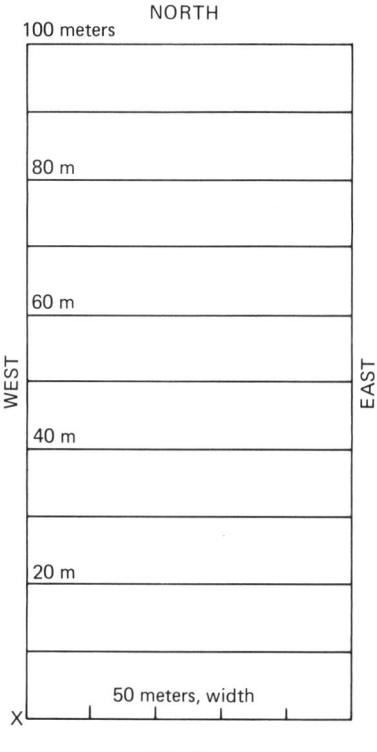

Physics Quiz

For each graph shown here, sketch the related curve to the right. All v t curves are straight lines.

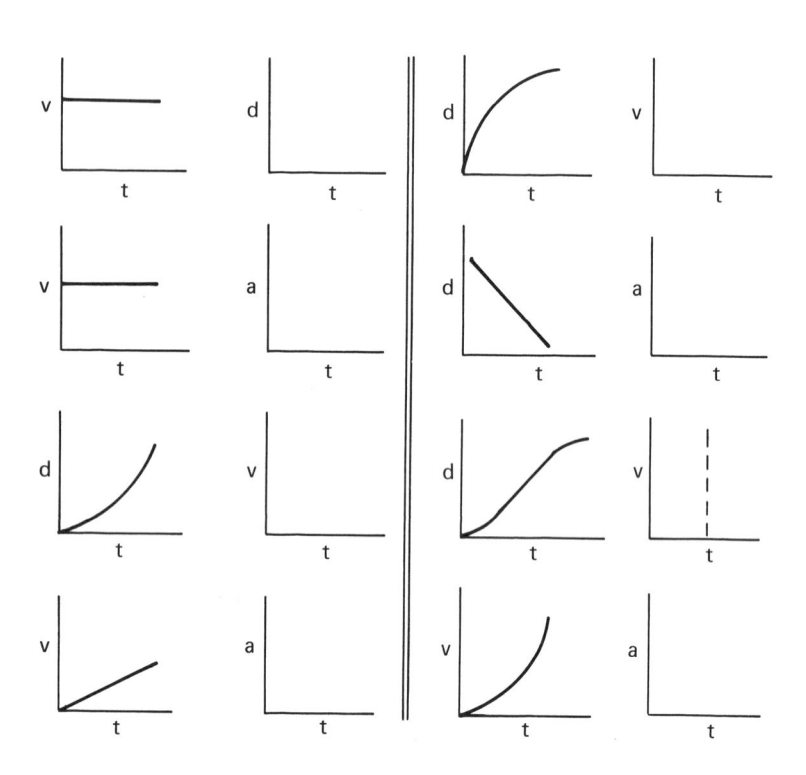

Quiz on Mechanics

1. I threw a ball from the ground and it came back down in eight seconds. How far did it go? Show mathematical calculations.
2. Analysis of Data:
   Calculate all items at the four-second point. What is the Acceleration?

Time traveled	Distance traveled	Velocity at end of time period	Average velocity per second
0 sec	0 meters	0 m/sec	0 m/sec
1 sec	1.33 meters	2.66 m/sec	1.33 m/sec
2 sec	5.32 meters	5.32 m/sec	3.99 m/sec
3 sec	11.97 meters	7.98 m/sec	6.65 m/sec
4 sec	_____ meters	_____ m/sec	_____ m/sec

Show calculations on a separate paper.

Test Review: An object starting from rest has a constant acceleration rate of 9 meters per second per second. At the end of this acceleration time, three seconds, it has zero acceleration for three more seconds. After this time period, it comes to rest (zero velocity) in six seconds, at constant acceleration (deceleration) rate.

Calculate the following:

1. Average velocity for the first three seconds.
2. Distance traveled in the first six seconds.
3. Acceleration rate from times: six-second period to twelve seconds.
4. Draw a velocity time graph for the total time interval.
5. Draw a distance time graph for the total time interval.

Take-home Problem

A ten-kilogram mass on the moon was dropped from above for five seconds. Put down all the related numbers, for every second, such as the calculations related to those of the Rational Newtonian Physics chapter.

Distance traveled, in meters, in one second						
Time (sec.) 0	1	2	3	4	5
Final Velocity (m/sec)						
Initial velocity (m/sec) List this for first, second and third seconds						
Acceleration, (m/sec^2)						
Force ma (newtons)						
Average Velocity (m/sec)						
Total distance traveled, in one second						
Total E_k, (j)						
Total E_p, (j)						
Work = **Fd** in one second						

Moon Acceleration = 1.5 m/sec^2

Use your own sources, but work this out on your own. You may have another in class one, related to this example.

Newtonian Test

1. An object goes forward from the rest. It travels two meters in one second and continues at this constant rate of distance increase per second, time, for four seconds. The line below represents this. Make and fill in all the blanks below that are related to specific time points and between specific indicated time points.

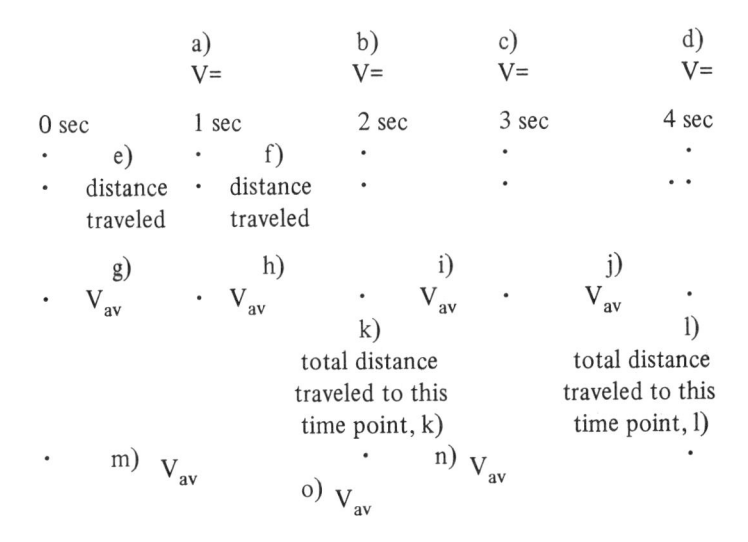

```
              a)           b)           c)           d)
              V=           V=           V=           V=

0 sec        1 sec        2 sec        3 sec        4 sec
  ·            ·            ·            ·            ·
    e)           f)                                    ·
  distance   · distance
  traveled     traveled

     g)          h)            i)             j)
  ·  V_av    ·  V_av      ·    V_av     ·     V_av     ·
                              k)                        l)
                         total distance            total distance
                         traveled to this          traveled to this
                         time point, k)            time point, l)
  ·   m) V_av              ·     n) V_av                ·
                       o) V_av
```

p) Calculate the acceleration rate.

"Blanks" above equal a) _____ b) _____ c) _____ to p) _____.

2. A one-gram-mass object 1000 meters up from the surface of a planet is dropped freely (due to the planet's gravity). In the first second it goes 40 meters.
   a) How long will it take to hit the planet? Clearly show how you calculate this answer.
   b) Is this planet bigger or smaller than the Earth? Explain your answer.

3. A ball dropped from the top of this classroom (ceiling) hits the floor in 0.75 seconds. How high is the room, floor to ceiling? Show your calculations for this answer.
4. A one kilogram ball is "shot" straight forward (horizontally) in the air at a constant velocity of 10 m/sec and lands on the ground in 20 seconds. (Ignore any friction in the air.) Calculate (and show your work) the following:
   a) The maximum distance it can go, forward.
   b) How high up it was from the ground.
   c) Diagram this vectorially, showing where the ball is every 0.5 seconds.
5. If the ball in question 4, were fired up from the ground (at 10 m/sec^2 speed, at the end of the first second), what is the maximum
   a) Height it can go?
   b) Distance it can go? (Calculate and show your work.)

Take-home Test Question

Open-book, if needed; individual work—honor system.

These questions represent many of your practice questions *and* review for the term test items. The best approach is to try to work out the answers *without* using a book or other references. This will show you how far you can get for very similar term test questions.

These questions allow you to use any book and notes, however, seek no help from fellow learners or anyone else.

Suggestion: Look over all the questions and think out what you already know. Then, work out as many as you can without book references. Finally, take each one and use as much reference as needed. The amount of references used will tell you how much you must go through for final testing.

Write answers carefully, completely, and clearly on the paper. Write out everything first.

1. A boy can throw a baseball horizontally with a speed of 20 m/sec. If he performs this feat in a convertible car that is moving at 30 m/sec

in a direction perpendicular to the direction in which he is throwing (90 degrees, throwing out the side window), what will be the actual speed and direction of motion of the baseball?

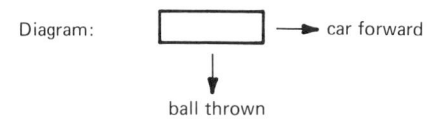

Diagram:

```
┌─────────┐  ──▶ car forward
│         │
└────┬────┘
     │
     ▼
 ball thrown
```

2. A car starts from rest and reaches a speed of 30 mi/hr in 8 seconds. What is its acceleration?
3. During the takeoff roll, a Boeing 747 jumbo jet accelerates 4 m/sec^2. If it requires 40 seconds to reach takeoff speed, determine the takeoff speed and how far the jet travels on the ground.
4. a) With what speed must a ball be thrown directly upward so that it remains in the air for 10 seconds?
   b) What will be its speed when it hits the ground?
   c) How high does the ball rise?
5. An object of mass 100 grams is at rest until a net force of 2000 dynes is applied for 10 seconds. What is the final velocity? How far will the object have moved in the ten-second interval?
6. What is the centripetal force required to cause a 3200-pound car to go around a curve of an 880-foot radius at a speed of 60 mph?
   Answer: 880 pounds. Show the calculations for this answer.
7. The mass of a bullet is two grams and its velocity is 30,000 centimeters per second. What is its kinetic energy?
8. A 40-gram bullet is fire from a 5.0-kilogram gun with a speed of 600 m/sec. What is the speed of recoil of the gun?

Essay Question—The Physics of a Home Run

Write this out carefully. First, answer the three questions with physics, and then write an essay on the basis of idea.

1. Do the home run hitters bat with straight or bent arms? Explain.

2. Is the stride as the home run hitter swings forward or backward related to weight of the body? Explain.
3. The home run experts start to swing when the ball is only 24 feet away, compared to 33 feet from the home plate, for the average hitter. The ball travels about 80 miles an hour from the pitcher. Explain this.

In terms of basic physics the following ideas are associated with a thorough scientific analysis of the feat: the speed of the pitch; distance of the pitcher from batter; his weight, diameter and resilience of the ball; length, weight, and composition of bat; height, weight and proportions of batter; speed and angle of swing; distance and direction the ball was hit; the arc it described; velocity and direction of the wind; and other factors.

This is the basis for the essay; put together a number of these ideas, physically. Try to design your own home run.

Term Exam

1. A two-kilogram mass starting from *rest* goes forward at a constant acceleration rate of 10 m/sec^2 for ten seconds. For the following related questions, show your calculations.

   A. What is the final velocity at the ten-second point?
   B. How far does it travel in ten seconds?
   C. What is the force applied?
   D. What is the average velocity, V_{av}, for the first seconds?
   E. If a 40-newton force is applied in the opposite direction (from its forward direction), how long will it take to stop the mass?
   F. Is the above statements vectorial? Explain.

2. You are in an elevator with no windows, and when you push lighty on the floor with your feet, you float up to the ceiling. Analyze and explain the direction in which the elevator is going, and as many Newtonian aspects as you can.

3. Consider the laboratory experiment where a mass attached to a string pulled down and made a cart move forward. The cart is a one-kilogram mass. The pulling mass is two kilograms. Assume that no friction exists in the system. At what rate will the cart accelerate? Explain by calculations.
4. A ship sails 10 miles in a northeast direction. What are the components of its displacement:

    A. To the north?   B. To the east?

5. A force of 5 newtons gives a mass, m_1, an acceleration of 8 m/sec^2, and a mass, m_2, an acceleration of 24 m/sec^2. What acceleration would it give the two when they are fastened together? (Show all work.)
6. Relate the speed of sound to your knowledge of the speed of light. What could sound be? This relates to the table below.

The *Speed of Sound* in Various Media

Medium	Speed (m/sec)
Air	332
Carbon Dioxide	259
Chlorine	206
Water	1404
Paraffin	1304
Copper	3560
Iron	5130

7. Look at an overhead projector demonstration. Describe how and why light makes this "picture."

8. A demonstration of sound phenomena. Listen carefully, then explain and analyze, relating the model of light to sound.

9. Graphs for analysis. Answer questions A through F.

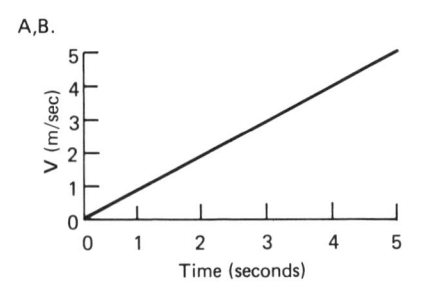

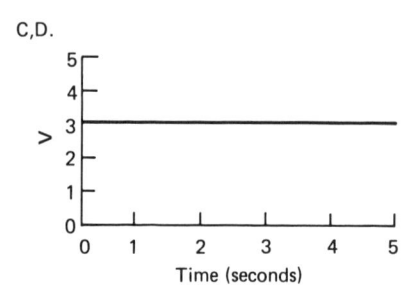

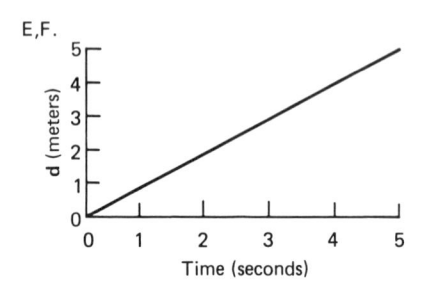

A,B.

C,D.

E,F.

A. What is the acceleration rate?
B. What is the distance traveled in 5 sec?
C. What is the acceleration rate?
D. What is the distance traveled in 5 sec?
E. What is the velocity?
F. What is the acceleration rate?

10. The chairman demonstration. Try to give a logical answer. Explain how it is constructed and what light phenomena are demonstrated experimentally.
11. Using the information shown on the velocity time graph below, carefully plot on **A** and **B**:

   A. A graph of distance/time for points A, B, C, and D. (Note: there is no "negative" distance.)
   B. A graph of acceleration/time for points A, B, C, and D.

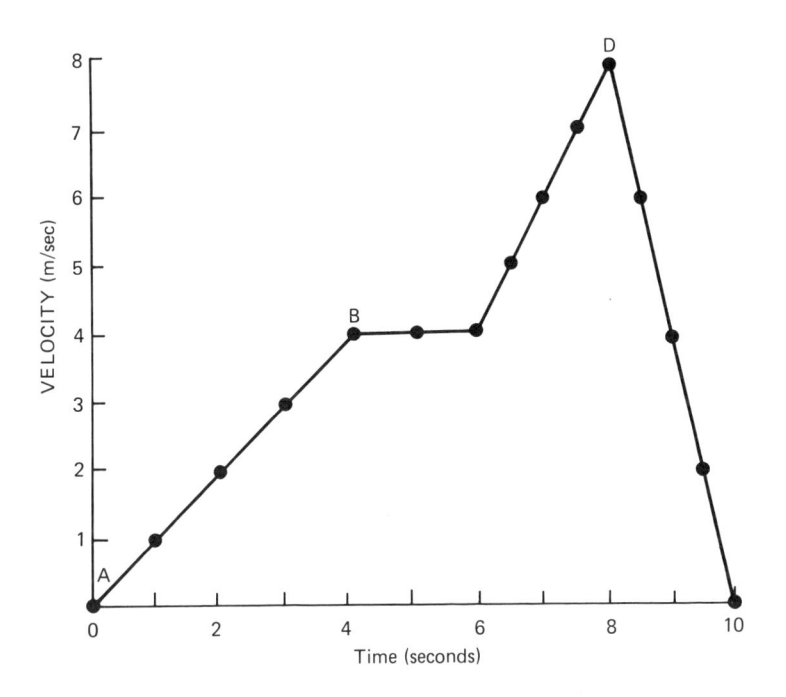

Time Graphs for Question 11.

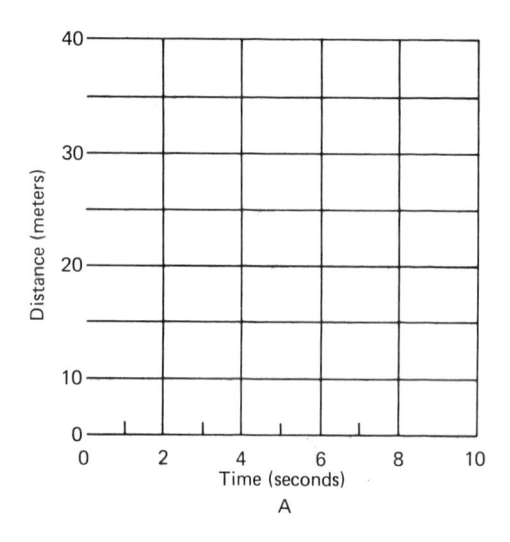

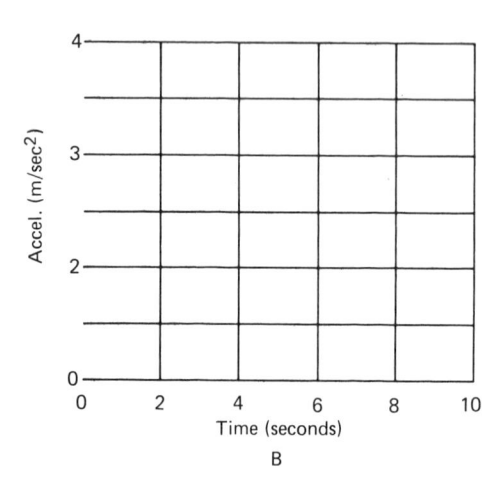

A

B

Oral Test on Newtonian Physics

Here are items for review that will be covered on the oral test among three or four students and a teacher.

I. An essay, presented orally, on the question What is Energy?
The group of four students should prepare to (1) talk about what energy is, initially in a general sense, then scientifically. Then, they should show how it is developed as a Newtonian concept related to all of the concepts that were already explained—mass, time, distance, velocity, speed, acceleration, force, work, momentum, inertia, and vectors. Finally, consider how energy should relate to further information such as heat, electricity, nuclear, and atomic reactions; and how it provides a better understanding of our world.

Be prepared to demonstrate all of these concepts in a sound mathematical sense, with good illustrations, and some simple physical illustrations. This essay should take at least a half hour to finish.

II. A number of demonstrations will be done that must be analyzed by the group.

On the essay (I) another teacher may interject questions that he considers appropriate and significant.

III. Questions on the Summary of Motion:
The teacher will state a number of basics to be put together, analytically. Some practical illustrations for analysis are car speed, rifle shooting, sports relationships, human utilization of energy, and others.

IV. Some simple questions on the parts learned:
What is X and how is it used to understand Newtonian ideas?

X may be velocity, acceleration, momentum, force, or another concept.

Under what condition(s) for a given system would you have the greatest PE, KE, velocity, acceleration, mass, or time?

V. What is conservation? Under what conditions (equations and physical phenomena) does it work, and when does it not work?

VI. Newtonian review questions, see the last page of "Rational Newtonian Physics."
VII. Related to this information, analyze how scientific laws grow out of observations, concepts, and hypotheses. Take a common physical observation and develop it as associated with:

   a. your knowledge of contemporary science
   b. a personal experience or observation
   c. any hunch that the problem may present to your mind
   d. simple quantitative relations or foreknowledge of the solution.
   e. new and interesting problems that are related to already known specific concepts
   f. Develop a working hypothesis on the observation. (This is associated with all the directions given in your oral lab reports.)
   g. Also, explain the *limitations* of ideas and relationships that can be gleaned from the observation, materials, and data results on hand. How much can you deduce other ideas?
   h. Design other experimental approaches that will progressively increase insight, and modify physical hypotheses.

Final Examination

This examination includes other science topics as well as mechanics. It is developed for present and later education. The many questions should give an analysis of your learning steps.

Multiple Choice Questions

1. Two vectors F_1 and F_2 act as shown at the right. Which answer below best represents the sum (resultant) of the two vectors?

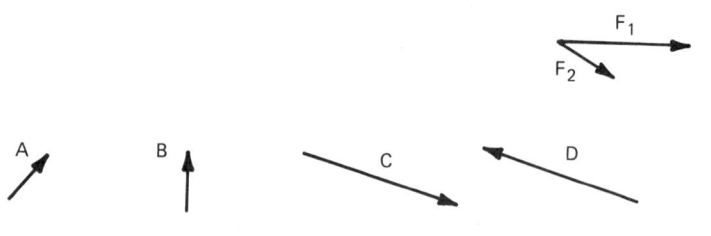

2. A distance/time graph for a cart moving with constant velocity will always be:

   A. a straight line parallel to the time axis
   B. a curved line
   C. a line with negative (downward) slope
   D. a line with positive (upward) slope

3. The slope of a distance/time graph enables us to calculate

   A. displacement   B. speed   C. distance   D. time

4,5. The following data for a car trip is to be used to answer 4 and 5.

Time Interval	Interval Duration (hours)	Average Speed (km/hr)
first	0.10	40
second	0.40	120
third	0.20	40

4. How far did the car travel in the first interval?

   A. 40 km   B. 4 km   C. 400 km   D. 0.4 km

5. What is the total distance traveled (in 0.70 hours)?

   A. 60 km   B. 140 km   C. 28.6 km   D. 40 km

6. Which graph represents the motion of an object moving backwards (toward the reference point)? (s = distance)

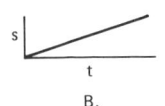

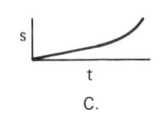

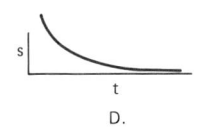

7. A cart starts from rest and accelerates uniformly to a speed of 8 m/sec in 2 seconds. The acceleration is:

   A. 0.25 m/sec^2 B. 10 m/sec^2 C. 16 m/sec^2 D. 4 m/sec^2

8. An inclined plane 10 meters long is shown below. Starting from rest at the top, a box weighing 100 nt accelerates at a rate of 2.5 m/sec². How long will it take the box to reach the bottom of the incline (10 m)?

   A. 2.8 sec   B. 2.0 sec   C. 4.6 sec   D. 4.0 sec

9. The unit of electrical power is the

   A. ampere   B. kilowatt-hour   C. volt   D. watt

10. When one gram of ice at 0°C has absorbed 100 calories of heat energy, the temperature of the resulting water is

    A. 0°C   B. 4°C   C. 20°C   D. 100°C

11. If a small nail is attached to a negatively charged rod, we can sure that the nail is

    A. magnetized   B. unmagnetized   C. charged negatively
    D. either unchanged or changed positively

12. Momentum is the product of

    A. force and distance   B. force and velocity
    C. mass and acceleration   D. mass and velocity

13. The number of calories required to melt 10 grams of ice at 0°C is

    A. 10   B. 800   C. 5400   D. 7200

14. If the resistance of a circuit is doubled while the applied e.m.f. remains the same, the current will

    A. be doubled B. be halved C. be quadrupled D. remain the same

15. The diameter of a lead pencil is approximately six

    A. centimeters   B. kilometers   C. meters   D. millimeters

16. When 10 grams of ice changes to water at 50° c, the quantity of heat absorbed by the ice and water is about

    A. 10 calories   B. 800 calories   C. 500 calories   D. 1300 calories

17. The leaves of a negatively charged electroscope diverged when an electrified body was brought near it. The electrified body must have been

    A. a conductor   B. an insulator   C. negatively charged   D. positively charged

18. Resistors of 10 ohms and 30 ohms are connected in series to a 120-volt circuit. The current flowing in the 30-ohm resistor is

    A. 12 amperes   B. 8 amps   C. 3 amps   D. 4 amps

19. The resultant of two 80-gram forces will be 80 grams when the angle between original forces is

    A. 30°   B. 60°   C. 90°   D. 120°

20. Which of the following is *not* a vector quantity?

    A. wind force of 300 nt on a sail   B. potential energy of 1800 ft-lb
    C. velocity of 40 m/sec   D. weight of 150 nt

21. The quantity of charge that is stored in a capacitor is measured in

    A. amperes   B. volts   C. farads   D. coulombs

22. If the resistance of an electrical circuit is halved and the voltage applied to the circuit is doubled, the current in the circuit would

    A. be quadrupled   B. be doubled   C. remain the same   D. be halved

23. When a bullet is shot from a gun, both the bullet and the gun acquire the same

    A. acceleration   B. energy   C. momentum   D. velocity

24. The temperature of 0.25 kg of water is raised from 22°C to 62°C in five minutes. The rate at which heat was added to the water was

    A. 1 kcal/min   B. 2 kcal/min   c. 5 kcal/min   D. 10 kcal/min

25. If equal masses of a metal and water are heated at the same rate, the temperature of the water would increase

    A. faster   B. slower   C. at the same rate   D. not at all

26. Uncharged spheres are attracted to charged spheres because

    A. charged bodies have an excess of one kind of charge
    B. charged bodies have a deficiency of one type of charge
    C. uncharged bodies influence charged bodies
    D. charged bodies produce an uneven charge of distribution on uncharged bodies

27. The potential energy of a system that consists of similar spheres with the same type of charge will be

    A. highest when the spheres are close together
    B. highest when the spheres are far apart
    C. lowest when the spheres are close together
    D. lowest when the spheres are touching one another

28. Work is measured in

    A. newtons   B. joules   C. coulombs   D. kilocalories

29. A positively charged rod is brought near a ball that is suspended by a thread. The ball is attracted by the rod, which indicates that:

    A. the ball must have a positive charge
    B. the ball must have a negative charge
    C. the ball is a poor conductor
    D. The ball may be electrically neutral.

30. Only one of two identical objects is electrically charged. Under which circumstances will the force of attraction be greatest between the two objects?

A. When the objects are touching each other
B. When the objects are close together but not touching
C. When the objects are widely separated
D. When the objects are touched together and then separated.

31. An inverse square relationship between two variables (X and Y) is best represented by which graph?

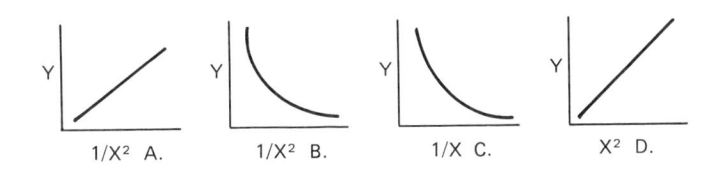

$1/X^2$ A. $1/X^2$ B. $1/X$ C. X^2 D.

32. An object with an excess of electrons is touched to the ball of an electroscope and then removed. The two leaves of the electroscope will then:

A. Stand apart because they have a positive charge
B. Be attracted together because they have the same charge
C. Stand apart because they are more negative than the ball
D. Stand apart because the system is negatively charged and non-polar.

33. A positively charged rod is brought near the terminal of a charged electroscope and the leaves collapse as the rod approaches the terminal. As the rod is brought still closer (but not touching), the leaves diverge again. This indicates that the electroscope was

A. positively charged   B. negatively charged
C. neutral   D. polluted

34. Uncharged spheres are attracted to charged spheres because:

A. Charged bodies have an excess of one kind of charge.
B. Charged bodies have a deficiency of one type of charge.
C. Uncharged bodies influence charged bodies.

D. Charged bodies produce an uneven charge distribution on uncharged bodies.

35. The electrostatic force between two like-charged bodies:

    A. increases as the bodies are moved farther apart.
    B. remain the same as the bodies are moved closer together.
    C. decreases as the bodies are moved closer together.
    D. increases as the bodies are moved closer together.

36. According to Coulomb's Law, the force between two charged bodies is proportional to which of the following?

    A. $\dfrac{Qq}{d}$ B. Qqd^2 C. $\dfrac{Q\ q}{d^2}$ D. $\dfrac{d}{Q\ q}$

37. The force between two charged objects is 100 newtons when the bodies are separated by a distance of 4 meters. What is the force when the same objects are separated by a distance of 8 meters?

    A. 50 newtons   B. 200 newtons   C. 25 newtons   D. 75 newtons

38. The potential energy of a system that consists of similar spheres with the same type charge will be:

    A. highest when the spheres are close together.
    B. highest when the spheres are far apart.
    C. lowest when the spheres are close together.
    D. lowest when the spheres are touching one another.

39. The potential energy of a system that consists of similar spheres with *unlike* charges would be:

    A. highest when the spheres are touching one another.
    B. lowest when the spheres are far apart.
    C. highest when the spheres are far apart.
    D. highest when the spheres are close together.

40. Two spheres are 1 meter apart. One has a charge of +1 charge units, the other −1. If the charges are increased to +2 and −2, the force between them will be how many times the original force?

A. 4 times the original   B. 1/4 the original
C. 1/2 the original   D. 1/16 the original

41. A charge of +1 is separated from a charge of -1 by an original distance of 0.1 meter. The original force is 100 force units. If the separation distance is changed to 0.4 meters, by how much is the force changed?

    A. 4 times the original   B. 1.4 the original
    C. 1/2 the original   D. 1.16 the original

42. The best explanation for the fact that a jet of water can be deflected by a charged rubber rod is:

    A. Water molecules are charged.
    B. Water is attracted by rubber.
    C. Water is electrically neutral but bears an uneven electric charge distribution.
    D. Water absorbs the charge from the rubber rod.

43-53. Write on the line at the right of each number in your answer sheet, the term that, when inserted in the blank, will make the statement true.

43. A man uses a force of 30 pounds to move a 200-pound object a distance of 5 feet. The work done is ___ foot-pounds.
44. A bomb is dropped from an airplane in level flight. At the end of 4 seconds the bomb has fallen ___ meters.
45. When the neutral body ___ electrons, it acquires a positive charge of electricity.
46. Five calories are required to raise the temperature of one gram of a substance 10°C. The specific heat is ___.
47. A 4-ohm resistor and an 8-ohm resistor are connected in series to a battery. If the current through the 4-ohm resistor is 2 amps, the current through the 8-ohm resistor is ___ amperes.
48. In order to attract electrons, the plate of vacuum tube should be charged ___.
49. A temperature of 10°C is equivalent to ___° absolute.
50. After falling freely for 3 seconds, a body that started from rest has a velocity of _____ m/sec.

51. A 50-kg boy climbs a 6-meter rope in 10 seconds. The work done is _____ nt-m.
52. A ball thrown directly upward with a velocity of 196 m/sec will reach its highest point in _____ seconds.
53. If a block of ice at 0°C is absorbing heat from its surroundings at a rate of 400 calories per minute, then the ice must be melting at a rate of _____ grams per minute.

In some of the following statements the italicized term makes the statement incorrect. For each *incorrect* statement, write the term that must be substituted for the underlined term to make the statement correct. For each *correct* statement, write the word *true* on the answer sheet.

54. A 2-volt, 5-ampere lamp and a 5-volt, 2-ampere lamp have the same *resistance*.
55. If the temperature remains the same, then the volume of a gas varies *inversely* with the pressure.
56. The smallest resultant that can be obtained by combining a 5 nt force and a 7 nt force is *zero* nt.
57. When a bullet is shot from a gun, the force acting on the bullet is *greater* than the force acting on the gun.
58. Heat is liberated when the stem in a radiator *condenses*.
59. Three calories will raise the temperature of three grams of water *three* degree(s) centigrade.
60. A body acquires a positive charge by losing *protons*.
61. When 1 cubic centimeter of ice at 0°C melts, the volume of the water produced is *less* than 1 cubic centimeter.

Problems
(show your setup and answer)

1. A charge of +500 esu is separated from a charge of –200 esu by 40 centimeters. Find the magnitude and direction of the field intensity at a point halfway between the two charges.
2. What current does a 100-watt lamp draw when connected into the original 110-volt house circuit? What is the resistance of the lamp?
3. How much current does a 600-watt electric toaster draw when connected across a 110-volt line? What is its resistance?

15

Newtonian Laboratory Items —Eleven Experiments

These laboratory items are mostly comprised of experiments and discussions with your teacher. It is a display of many steps that relate to most of the information that was explained in this book. Almost all physics topics are related to your human, individual obervations. Having looked at or handled a number of physics steps, a student can come up with a very personal scientific analysis. Try to develop good answers individually —and this should start to become a lot of fun, which is helpful for good learning.

Some of the items are a "step forward" from mechanics, to bring out the student's potential for understanding the following steps.

Analysis of a Dropping Egg

Your teacher will drop two eggs (or more) and only one will break. Scientifically observe this phenomena, and then try to develop good answers, which should relate to a true understanding of practical living. This relates to physical items. Related values are its mass, weight, distance/energy, and the surroundings.

In the egg experiment, one fun step could be for two students to start tossing an egg back and forth. The boy (or girl) can catch it without breaking it, as the students go farther and farther apart. The pair that is farthest apart without a broken egg is the winner. It all relates to mass, distance, time, and velocity. An important aspect of science relates to the "science of people," such as steps like the egg experiment.

Basic Colors

Place a red paper (about 8 inches square) on a larger white paper. Now, stare at the red item on white paper, until teacher tells you to stop. Then, look at the white paper; what color do you see? Explain this. The view, that you will see relates to our head and brain physiology. You will also look at blue and green papers, the same way.

The variation of colors is a very basic aspect of physics. It relates to your individual knowledge, for really understanding things as you see them. A practical question is: when one buys clothing of a specific kind of color the color may be wrong. Why?

Shooting—Equal Mass and Weight

The shooting equipment has the same force every time. Shooting should be straight forward, then at angles up to 15 degrees, 20 degrees, 40, and so on. Which bullet will go the farthest, and why?

This demonstration relates to practical steps; when you play baseball, at what angle can the ball be hit such that it will go the farthest? Explain.

This can also be associated with your "brain shooting." Which of your personal learning steps provides the best knowledge?

Different Masses Knocked Forward

A heavier end-item is to be fastened onto the above setup for shooting like the former demonstation. Watch the hit, at equal angles. How far does the heavier one go, related to the lighter one? Different masses that are hit by equal energy will go forward. How far does each go, and why? With such a lab there are many mechanical physical items to be put together.

Another practical aspect is that different masses relate to some of our great needs today, such as big versus little cars. If both travel at 55 miles per hour, which needs the most energy? If they have an equal amount of energy, which one will go the fastest?

Candle Quencher

Some wire has been wrapped around a pencil to form a coil. For the demonstration, slowly lower this coil around a candle flame. Then, write your observations. How does the wire coil reduce or extinguish the candle flame?

Man, Seat, and Man in Seat

For this experiment, see the illustration below. Turning on light one, you will see a seat, and light two, will show a man. Put the lights on together and note what you have seen. How does it work?

A covered wooden container, about 5 inches high from bottom

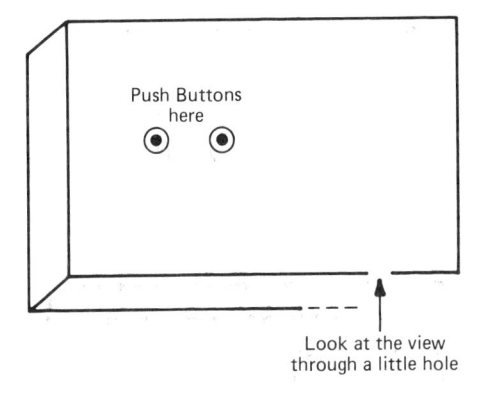

With this experiment you may not be given a real answer. It relates to light and it's relationships under certain conditions. Many students develop a significant diagram for this, often different from the "real" BUT equally as good!

Jim Dandy X-Ray Machine

Diagram

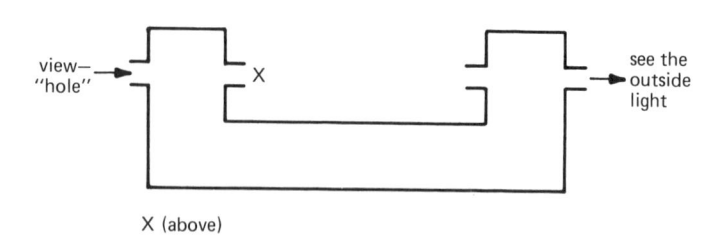

```
view—        ┌─┐         ┌─┐     see the
"hole"  →    │ │    X    │ │  →  outside
             │ └─────────┘ │     light
             │             │
             └─────────────┘
```

X (above)

Look through this item, which has one hole, and point it at a light that you can see. Then put your hand in the middle hole's area to cut out the light. What happens, and why? Explain the structure of this equipment.

Army Man Walking

A picture shows a number of army men all walking forward together at the same rate. They are first on a solid street. Then, they walk into an area away from the street, a "sticky," hard-to-walk area. Note that the group turns in a different direction. Explain.

This relates to some personal views and ideas that are very much related to the whole world. An associated question is: When looking at a fish in the water, what kind of angle are you looking from? Why?

Titanium Dioxide

Let a first letter be blue and a second letter be red. Look at the two letters together through a round glass item with water in it. Look at it closely, then farther away three or four inches. Note what you see, and

Explain. This is a truly simple, and fun, knowledge step. It relates science to other academic areas.

What Keeps an Airplane in the Air?

For this demonstration you may be asked to develop a laboratory item by yourself, or write an analysis. It relates to forces on an airplane, and how the air that flies through the area keeps the plane up. What does this tell us about air on an airplane?

House Energy

This is a scientific idea, that is useful for the future. This item shows once again how science can be related to very practical knowledge. Energy is needed to put something together such as a house. It usually takes months to build this house. Then, the house exploded in *seconds* and came completely apart. How do you relate these two energies?